An Expanding Theology

Tony Kelly is an Australian Redemptorist, whose studies include doctoral work in Rome (STD, Anselmianum) and post-doctoral studies in Paris, Toronto and other centres. He has lectured extensively in theology and has been president of the Yarra Theological Union, Melbourne, and the Melbourne College of Divinity. His publications include ten books, among them *Touching on the Infinite: Explorations in Christian Hope* and *A New Imagining*, and numerous articles in journals such as *Theology Review* and *Theological Studies*.

Today's discoveries in cosmology and ecology are expanding our understanding of the universe in an explosive way. This poses a challenge to theology to express the faith in new ways, within the narrative of the expanding universe and the emergence of life on this planet. With a sure touch Tony Kelly brings the new science face to face with ancient Christian teachings about Christ, creation, anthropology, Trinity, Eucharist, death and love. His language is accessible, his enthusiasm is infectious, and his insights are profound. This is a beautiful book on a subject of the greatest importance.

Elizabeth A. Johnson,
author and Professor of Theology,
Fordham University, New York.

One of the perennial difficulties faced by a contemporary theologian is the need to draw from the richness of Christian tradition while facing the challenges of a radically different and rapidly changing world which must be addressed. Tony Kelly has done this magnificently. I wonder at his wide-ranging interests and knowledge: from classical to contemporary theology to poetry to metaphysics to modern scientific research to human sexuality and death. What is more wonderful is his passionate commitment to an integrated presentation of them all. This is a fine and important book. I hope that it is widely read.

Francis J. Moloney, SDB,
author and member of the International Theological Commission,
Professor of New Testament Studies,
Catholic Theological College, Melbourne.

Tony Kelly has painted a large impressionist landscape of creation as a place of divine and human delight. Guiding the colour and verve in his writing is a commitment both to the riches of the Christian intellectual tradition and to the wonderful revelations uncovered by our restless human spirit in literature, philosophy and science. This brave book offers a pattern for theologies of the future.

John Honner, SJ,
author and lecturer in theology and the philosophy of science,
Jesuit Theological College, Melbourne.

AN EXPANDING THEOLOGY

Faith in a world of connections

TONY · KELLY

E. J. Dwyer

First published in 1993 by
E.J. Dwyer (Australia) Pty Ltd
3/32–72 Alice Street
Newtown NSW 2042
Australia
Phone: (02) 550-2355
Fax: (02) 519-3218

National Library of Australia
Cataloguing-in-Publication data

Kelly, Anthony, 1938–
An expanding theology : faith in a world of connections.

Bibliography.
ISBN 0 85574 229 1.

1. Theology, Doctrinal. 2. Faith. I. Title.

230

Cover design by Todd Davidson
Text design by Katrina Rendell
Typeset by Egan-Reid Ltd, NZ
Printed by Southwood Press, Marrickville, Australia

Distributed in Canada by:

Meakin and Associates
Unit 17
81 Auriga Drive
NEPEAN, ONT K2E 7Y5
Ph: (613) 226 4381
Fax: (613) 226 1687

Distributed in Ireland and the U.K. by:

Columba Book Service
93 The Rise
Mount Merrion
BLACKROCK CO. DUBLIN
Ph: (01) 283 2954
Fax: (01) 288 3770

Distributed in the United States by:

Morehouse Publishing
871 Ethan Allen Highway
RIDGEFIELD CT 06877
Ph: (203) 431 3927
Fax: (203) 431 3964

Acknowledgements

The author gratefully acknowledges the use of material from the following works. Every effort has been made to locate the sources of quoted material and to obtain authority for its use:

Extracts from 'Religions are Poems' by Les Murray from *Blocks and Tackles*, © Les Murray, 1990. With permission from Collins/Angus & Robertson Publishers, Pymble, NSW, 1990.

Extracts from 'The Forest', 'Five Senses' and 'Connections' by Judith Wright from *A Human Pattern*, © Judith Wright, 1990. With permission from Collins/Angus & Robertson Publishers, Pymble, NSW, 1990.

The Glory of the Lord, Hans Urs von Balthasar, T & T Clark Ltd, Edinburgh, 1982.

Vatican Council II, Conciliar and Post-Conciliar Documents, Editor, Rev. Austin Flannery, O.P., Costello Publishing Company, Inc., Northport, NY11768, USA, 1975.

'Lives of a Cell' copyright © 1971 by the Massachusetts Medical Society, from *The Lives of a Cell*, by Lewis Thomas. Used by permission of Viking Penguin, a division of Penguin Books USA Inc.

The Philosophy of Social Ecology, Murray Bookchin, Black Rose Books, Montreal, 1990.

Method in Theology, Bernard Lonergan, Darton, Longman and Todd, London, 1972.

Excerpts on pages 88, 176 and 214 from *The Divine Milieu* by Pierre Teilhard de Chardin. Copyright © 1957 by Editions du Seuil. English translation copyright © 1960 by William Collins Sons & Co Ltd., London, and Harper & Row, Publishers, Inc., New York. Reprinted by permission of HarperCollins Publishers.

The Denial of Death by Ernest Becker. Copyright © 1973 by The Free Press, a Division of Macmillan, Inc. Reprinted with the permission of the publisher.

Scriptural citations are taken from *The New Revised Standard Version Bible*, Oxford University Press, New York, 1989.

CONTENTS

Preface

This book is about Christian faith making new connections; specifically in the areas of ecology and cosmology. Clearly, I will be dealing more with the background against which such connections might be made. Or, to vary the metaphor, I try to suggest the horizon in which a new sense of connectedness to the biosphere and the encompassing cosmic reality can come to expression.

Certainly, many Christians have an increasing familiarity with the brilliance of the scientific explorations that are taking place, all the way from quarks to quasars. This new sense of the mystery of the cosmos is often accompanied by a stirring of ecological conscience: they wonder that the universe has brought forth life in all its precious variety. In such an expansion of consciousness faith is temporarily tongue-tied. How can our Christian vision encompass the wonder and responsibility that a new sense of reality inspires? How can faith make, and live, these new connections?

Such are the questions I try to address. Of course, you can only do that from a limited point of view. There is no universal standpoint outside the whole process: you can only enter the conversation with what you are, and where you come from. Essential aspects of what I am and where I come from are, first of all, Christian faith, the Catholic tradition, and the discipline of theological exploration. More immediately, my viewpoint is affected by the degree to which I have let that faith, that tradition and that discipline be sparked into a new intensity of exploration by new holistic and cosmic perceptions of reality. Unless you are to become progressively schizophrenic, you have to try to make sense of it *all*, somehow. And that is where we touch the problem.

It requires no special modesty to admit that we do not know everything about everything. Neither do I think it is an excess of humility to disclaim knowing even something about everything. Still, I found myself, in the course of writing, often returning to one of the most enduring philosophical insights, from before Aristotle to after Teilhard, namely that each of us *is* something about everything, or at least, somebody in the everything—the microcosm within the macrocosm, the image of God within the universe of God's creation. To bring *that* to consciousness, to cultivate that background sense of belonging to the whole, to throw light on our existence as an overture to the All, to intimate the horizon in which we are participating in the emergence of the universe, struck me as a worthwhile goal.

Yet there is a risk in doing that. You have to work in terms of intimation, evocations, and a feel for creation as poem before you read it as a work of prose—general statements, in the most outlandish sense of the word. It might all sound irritatingly vague, not only because of my limited competence, but also because we are no longer much used to thinking in the largest possible terms.

Still, I am disposed to defend an amateur status in this new age of connections. To suggest that we are all amateurs when it comes to 'getting it all together' implies no arrogant judgment on the expertise and intelligence—be it in science, theology or philosophy—that have been exploring different dimensions of these cosmic and global matters. The universe as it emerges into our present apprehensions does not encourage professionals. It favors the amateur, in the two senses of the word. First, in that we are all beginners in the presence of the overwhelming complexity of things. Though some might have a remarkable competence in one or another or even several of the relevant disciplines, there is no room for the cosmic know-it-all. None of us can be so full of ourselves as to render further discussion useless.

Things are too big for that and, on the shadow side, too far gone if we dare face either the ecological destruction that has occurred in the world, or the distress occurring in the inner ecology of our various cultures.

But there is a second sense of 'amateur', an older and more tender one. In its Latin roots, it means quite simply, 'lover'—less romantically, one who does things for the joy of participating in something of essential value. This is a meaning I especially treasure in the present context. I think it captures a sense of not waiting for the grand solution, but of starting where we are and making the best of it, by throwing our best selves into the whole business. If you left love or art or communication to the professionals, we would hardly begin to live.

I think, too, that in the deep reaches of faith, such amateurism is most valued. There are no professionals in the sight of God; we are saved neither by the knowledge or the good works of someone else. We enter into the Kingdom as 'amateurs', as lovers, as eternal beginners. This is especially the case, perhaps, when it comes to making new connections with the wonder of life and the universe itself.

In due course, we can hope that refinements of intelligence and skill will find the right words and come up with the right methods; but such achievements would fall flat if there is no tide of love, no world of living connections, on which to ride. To that degree, a newly connected world has to be one of fresh democracy. There, everyone can vote with the authority of a particular love born of a special experience, however inexpressible, of universal connections.

Whilst the various subjects listed in the Table of Contents are predictable enough, I am writing with the purpose of nourishing or sparking a conversation. I would like to think that I am presenting the matters discussed in an ecology of meaning, with the accent on connections, implications and analogical linking—a particular version of that 'framework of collaborative creativity' that Bernard Lonergan commended in his *Method In Theology*. However, 'framework' is perhaps too much a metaphor of stability, given the more spiralling pattern of connections I have attempted to make.

This book has a beginning and an end; but at any number of places within it, the reader might find his or her own point of entry: the procedure I have adopted permits that, if only for the reason that I often puzzled over the best sequence myself. After all, one can drop in and out of a conversation: neither here, nor in most domains of life and growth, does linear logic claim a very tight rule.

I begin with two rather large 'circles of connections': the first, *Implications*, introduces the expanding scope of faith-seeking connections; the second, *Contexts*, concentrates on particular elements within that scope. I then go on to a series of other arcs of reflection. These are more thematically focused in Christian considerations: Christ, creation, anthropology, Trinity and eucharist.

It might be surprising to find that the whole thing ends with the two 'dimensions' of universal experience: death and sexual love. I wasn't quite sure where to treat such matters—but then one seldom is. All I know is that any seriously playful effort to make connections can hardly leave them out, even though I have found, to my surprise, that a good number of those who write on our new ecological or cosmic connections make no mention of such inescapable dimensions of life.

So, a spiral of meanings: whether this will appear as a harmless stylistic conceit on my part, or more seriously, a subterfuge to cloud the issue—perhaps to conceal my ignorance of hundreds of technical issues—or something more helpful, will appear in due course.

There is that risk I mentioned above, in trying to 'get it all together' in some way: you leave so much out; you merely touch on so much else; you make summary reference to what is already well done; you keep making the almost daily discovery of further key sources.

And you keep wondering who is likely to read this kind of thing. I suppose all writers have 'a general reader' in mind, usually a composite of the author's nightmarish fantasies: a team of experts poring over one's words with the diligence of scholars examining the Dead Sea Scrolls; or the 'simple faithful' aghast and dismayed; or 'lost souls', perhaps, finding only further reason for confusion and doubt.

But a sense of reality returns. When I remind myself of the conversations from which this book emerged, and those to which I hope it may lead, I find that experts are usually interested in a new angle, that the 'simple faithful' have a powerful good sense and that 'lost souls' are the great searchers. Perhaps there is something of all three in each of us these days.

You will notice, too, that I make some effort to blend new things and old. Though our sense of time, and of reality generally, is vastly different from that of the medievals, I am convinced that there is a great deal to be retrieved in the vision of Aquinas and his contemporaries. Part of the thinness of the modern theologies of ecology and creation result from trying to invent the world from scratch—a theological reinvention of the wheel, you might say—without taking account of past achievements.

Of course, I am indebted to many other theological influences, as the references will show. Pierre Teilhard de Chardin, Bernard Lonergan, Karl Rahner, Juan Luis Segundo, along with their various astute commentators and continuators, are always at my elbow.

How I have succeeded, what this book adds to the literature in this area, readers will judge for themselves without any assistance from the author. All I can do is once more to express the hope that any given conversation dealing with this universe of connections might be reminded of aspects that tend to be forgotten, and be inspired to a fresh exploration of what 'catholic' (*kata holou*, 'in accord with the whole') might mean today.

Every book is the outcome of a whole ecology of shared resources, care and conversation. I cannot begin to name all the good influences that have supported and challenged me in my writing in this instance. But I must mention some: first, Catherine Hammond of E. J. Dwyer joined energetic encouragement to meticulous editorship to make me say what I mean more clearly. Then, I always find myself thanking a very generous colleague, Michael Mason, C.Ss.R., not only for his personal support but for his technical assistance in producing the finished manuscript. My religious community were, as usual, patient with the strange preoccupations that possess someone writing a book in their midst. I thank them for their indulgence. Another Redemptorist confrere, Brian Johnstone, C.Ss.R., of the Alphonsian Academy in Rome, helped me, during his recent visit, with a variety of shrewd observations on certain sections of the manuscript. Finally, there was Gwen McGrenere of Toronto, Canada, who, some years ago, started a conversation with me that has ended in this book. Such names stand for a multitude. They bring home to me the reality of what I am here exploring, the world of gracious connections in which we live.

Tony Kelly, C.Ss.R.

SECTION I

A First Circle of Connections: Implications

1. A turning point for Christian theology

I believe that Christian theology, by expanding to meet the demands of the age, is coming into its own as a great intellectual adventure. As faith works to express its meaning in the light of new ways of understanding the universe and in the context of a new ecological awareness of our planetary coexistence, there are plenty of splendid beginnings. Yet, at the moment, we are still at the point of inklings, anticipations, partial viewpoints in regard to what might emerge. The task ahead is daunting. There is nothing new in that. As with all emerging realities, there will be deadends, strange turns, long periods of puzzlement, much confusion, discouragement and sudden breakthroughs, as the course of trial and error zigzags through the range of possibilities to come to decisive insights and, eventually, to a larger vision. The great medieval theologians knew the burden of such creativity as they struggled with the newly available writings of Aristotle, and pored over the Jewish and Arab commentaries on 'the philosopher', as Aquinas would always call him. But while the medieval achievement is both an inspiration and an often surprisingly fruitful resource for what is before us now, it leaves us, nearly eight centuries later, with distinctive challenges and opportunities.

As a reflective Christian faith slowly familiarises itself with the methods, the categories, the achievements of current explorations of the world, of life and of the cosmos itself, a definite excitement stirs in theological thinking. It is beginning to learn a new language, one more worthy of the limitless mystery it serves.

It seems we are coming out of a period in which the language of faith was a very private dialect, just as 'the light of faith' was often experienced as a very interior, inward illumination. There were reasons for this and history does impose its limits. But now that dialect has a chance to become a means of universal communication as it gives expression to a new sense of gracious wholeness. Likewise, the light of Christ has begun to play on the whole mystery of life and to irradiate the universal scope of our knowledge of the world. In short, faith has the opportunity to express itself within a new learning curve in the shared human search for the most comprehensive vision of the universe and of the genuinely earthly character of our existence.

Christian faith is natively a universal vision of all things in Christ. Hence it must always be looking for an adequate 'philosophy' to articulate its real meaning in terms of the 'all' and the fundamental meaning of the real.[1] When such a search for wisdom shares today's ecological optic and unfolds within an expanding cosmological horizon, something new is beginning. Theology is being reinvigorated with the excitement of learning. This has already happened in terms of the fundamental documents or doctrines of faith: enormous progress has been made in biblical and historical studies. But there remains 'the book of nature' to be read as a kind of primary revelation accessible to all.[2]

Then, too, the expression of faith has already achieved a new comprehension of social reality in the variety of liberation and political theologies. But such modes of reflection eventually give rise to deeper questions concerned with our coexistence with all living things on this planet. Also, there is the increasing expert practice of dialogue with other religious traditions and worldviews. But here, too, we must go further. While the growth in mutual comprehension and respect are magnificent achievements, neither our differences nor our agreements can let us forget our common earthly origins and responsibilities.

Some would see the feminist turn in religious and theological awareness as the decisive advance. But there is more still: while faith can be newly converted to those religious, intellectual, moral and psychological values embodied in the unacknowledged experience of women, such values invite us to a deeper, more inclusive sense of life and reality which remain to be explored.

In short, while genuine advances in the search of faith are secured in a more critical scrutiny of its primary data, in a more compassionate social praxis, in dialogue with all peoples of wisdom and goodwill, in the overcoming of sexism, each of these advances can only be enhanced when set within the movement of a longer, larger, all-inclusive story—that of the emergence of the universe itself and the evolution of life on this planet. When the Gospel story is told within such a narrative, faith begins to reach down to its deepest roots in the earth; and to stretch out into hitherto unimaginable dimensions of the cosmos.

2. Faith making connections

In this new context, faith (not to forget hope and love) continues 'to seek understanding', *fides quaerens intellectum* as the time-honored definition has it. Though the current context of this search is often overwhelming in its novelty, three basic and rather traditional techniques stand out as resources for a renewed elaboration of the Christian mystery.[3]

First, there is the way of analogy. By using models, metaphors, symbols drawn from experience, we move from the known to the unknown in order to arrive at a partial but more comprehensive expression of what faith means. This is the most familiar of all theological methods. Hence, we explore the meaning and significance of the Trinity, following the paths mapped out by, say, Augustine and Aquinas, as they extrapolated from the experience of human community or human consciousness.

Secondly, theology constructs its systems of meaning by making connections between all the different articles or aspects of faith to achieve an ordered, one might say holographic, vision of God's self-communication in Christ. For example, the mystery of the incarnation can shed light on the meaning of the sacraments; just as the eucharist, for instance, can suggest ways of understanding how the Spirit is present and active in the world.

Thirdly, theology seeks to speak to the dynamics of hope, and to the search for meaning and fulfillment. It must immerse its religious expression in the stream of humanity's common search for the fullness of life, the ultimate in human destiny. The meaning of faith is meant to be liberating, leading to a greater conduct of freedom, yet opening to that future that only God can give. The dramatic and often heroic articulations of Liberation Theology move in this direction.

In the present scientifically informed culture, such analogies, interconnections and ultimate reference need to operate in a field now incredibly extended in comparison with the past. An expanding universe, an evolutionary world, a relational and holistic understanding of all reality demand a theology that can grow with the universe of our perception. Reflective faith, therefore, has to work with analogies drawn, for example, from quantum mechanics and the evolutionary process.[4] The old views of nature, of matter, of substance, even of causality, will not do. The unchanging 'great chain of being' that structured the mental space of past thought for millennia must now be interlinked within the processes that have occurred, and are occurring, in time.

Perhaps the deepest change in the style of analogical thinking comes in the realization that the whole universe of reality is one great process of evolutionary emergence. While there are distinctions, hierarchies, differentiations to be observed in the given plurality of living and non-living things, all are alike inasmuch as they participate, in an interconnected manner, in the one process of universal becoming. The cosmos is understood as a vast analogical event. If analogy is knowledge made possible by extrapolating from one frame of reference to another, it is now seen, not so much as an improbable stretching of the imagination from one order of being to another, but as the only way of approximating to the concrete, dynamic unity in difference that the actual universe represents.[5]

As there are new resources for analogical thinking, so too there are new 'interconnections' to be made. The connections that faith might make amongst its own specific truths need now to be related to the world of interconnections and communication that science has both discovered and enabled.

If new analogies recommend themselves, if there are new connections to be made, there is also a new form of universal hope to be expressed. Faith has to learn to speak of human destiny not simply out of a laudable concern to save one's soul, but in terms of the emergence and destiny of the universe as a whole. Theology is faced with its most important question: how daring and inclusive might our hope be?

In all this, reflective faith has much to learn, and in that learning, something unique to share. The light of faith is shining in a radiant universe. We can perceive a universe literally aglow with cosmic radiation witnessing to its explosive beginnings. NASA's Cosmic Background Explorer (COBE) satellite is enabling scientists to backtrack fifteen billion years to the first 300,000 years of our cosmic beginnings.[6] In a world of such stupefying dimensions of time and space, the light of faith is another kind of background radiation, gently inviting us not to miss the whole point of our existence.

Thus, the 'faith seeking understanding' of theological tradition is struggling for new expression as 'faith making connections' with, and within, the ecological and cosmological understanding of our day.

3. The theological spectrum

The current situation of theology can perhaps be illumined by the simple metaphor of the spectrum of the rainbow. Up to a comparatively recent past, the great theological and philosophical traditions that have nourished the expression of Christian faith have tended to concentrate on one or the other end of the spectrum of our experience of the reality. At the upper extreme, theology explored the ethereal indigo of human transcendence and spirituality, and the violet of the inexpressible mystery into which it fades. More recently, reflection has been concerned with the lower extreme, the red of the flesh and blood of our humanity, as it is manifest in, say, various current efforts to affirm the body or the feminine, and to face more squarely the extent of human suffering. Consequently, it has moved into the dramatic orange and gold of essential human values prized as the animating force of human culture.

Between the two extremes of the spectrum is the green of a new ecological realization. In this middle band, the vivid colors of our history and culture come together with the blue of larger cosmic exploration, to meet in the hitherto unnoticed green. There it finds the earth and all its living systems in the great community of planetary life. This middle band of green, in its turn, unfolds into the blue of the cosmos, as into the indigo and violet of the unfinished reality that we are, and the ultimate mystery which our souls breathe.

Simple as this metaphor is, it is a way of saying that theology has begun to respond to all the colors of the spectrum. We no longer live in the old simplicities of black and white. For we find ourselves in an astonishingly vast, subtle and beautifully differentiated universe of many colors. With this new sensitivity to the variegated radiations of the light, faith can befriend the ultimate mystery which dwells in the radiance of 'unapproachable light' (1 Tm 6:16) of the divine mystery; and appreciate more fully the all-illuminating light which, in Christ, 'was coming into the world' (Jn 1:9).

As it unfolds into this larger, more differentiated arc of consciousness, a viridescent theology is emerging. Green has become, as it were, the linking color, blending the precious gold of human consciousness with the overarching blue of a larger cosmic mystery. Our whole experience of reality is undergoing a 'greening'—of the church, of the Gospel, of ethics and morality, of science, spirituality and culture—all converging to argue, one way or another, that 'God is green'.[7]

Metaphors aside, a new movement, a new connection, a new comprehension of the whole is taking place. Perhaps most of all, it is a new taste for life and the mystery of its origin and meaning. Faith has its own intimacy with that meaning and origin. For that reason the words of the psalm leap out and take wing in directions that theology has to learn ever anew to follow. It has to learn to speak to and, more importantly, out of, the awareness of those who, with growing realism, can

> ...feast on the abundance of your house,
> and you give them drink from the river of your delights.
> For with you is the fountain of life;
> in your light we see light (Ps 36:8f).

4. Faith in a time of complexity

Despite the fact that the ecological turn in our culture and faith seems irreversible and that the new cosmological horizon is one of limitless excitement, it is inevitable that a new range of monochrome ideologies will try to prevent the full play of possibilities. But a world radiating only green or red or blue light would eventually prove nauseous. Before long, we would be living in an anxious world of jaded, depressive gray, where the only realities are the problems, the conflicts, the failures and the threats.

It is not the time for faith to lose its nerve. For it contains vital resources to center and support the new turn in human awareness that has begun, but which remains vulnerable on many fronts. Even to be minimally touched by the new ecological awareness and the scale of the cosmos is to feel intimidated by the scope of the challenge. Though a certain amount of vertigo is inevitable and healthy, the newly wakened consciousness risks becoming intoxicated, overloaded, and finally exhausted. How to keep the vision alive and the contribution positive and vital, how to grow both in hope and the joy of existence, become real issues. After all, the scale and intensity of new global responsibilities can cause 'burn-out' of the same proportions!

For instance, as we begin to take in the proportions of the destruction that has occurred in the tiny span of recent human history, the negativities of the situation can be appalling. Add to this the tides of information that wash over our minds from the dozens of relevant sciences and the hundreds of activist groups, and we soon find that it is all too much. A meeting held in New York in preparation for the Earth Summit in Rio de Janeiro in June 1992 produced twenty-four million pages of documents![8]

Obviously there is a marvelous heightening and expansion of consciousness to be had as we think of the millions of species to which we are related, sometimes, alas, destructively, in the commonwealth of planetary life; as we ponder the billions of years that have gone into the making of this present moment of shared existence; as we grow in awareness of the beauty, the fragility, the sheer chanciness of it all. Yet there is a special strain as well. The imagination unfolds into strange new frontiers of experience. We are taken to the edge of a limitlessness beyond: the meaning of it all, the wonder of life, the obscure purposes that govern the whole of existence, the directions that beckon. Likewise, we come up against limitations within: how little we know and how little we can do.

To touch on such limits is to be driven back to the most sustaining sense of reality. We begin to feel a fresh need for some inclusive horizon, some solid ground on which to stand, some shareable source of energy and hope.

Aristotle averred twenty-three centuries ago that the human spirit was in some way all things.[9] But it is no easy matter, psychologically speaking, to be open to the universe, especially as contemporary sciences are revealing it. The new vision not only disorients the mind as any number of the pioneers of quantum physics confessed[10]; it disturbs the conscience as new values, new cares, new responsibilities make their presence felt.

Given the complexity of the situation, any inherited meanings and values suffer a crisis. They are stretched to breaking point. Any pretended synthesis leaves so much out. We soon find ourselves adrift in problems and brought to the point of a new kind of existential panic. The danger is that we either give up in hopeless confusion, or seek an escape into a rigid one-dimensional solution. At that point, a premature totalisation of one or another aspect of the problem or the answer blocks further learning. Psychologically, we cease to be participants immersed in the whole emerging reality. Our minds congeal in the midst of the stream of life. We no longer go with the flow.

It is then that imagination fails; and with it, the true patience of hope. We no longer have time for the whole to reveal itself. More practically, the possibilities of genuine conversation to be enlivened by the expanding range of human expertise and experience are diminished.[11]

Our century has known enough of the destructive force of totalitarian ideologies. The curse of any 'grand solution' is that it works against the very realities it pretends to cherish. For instance, a scientific ideology becomes a learned refusal to learn. A socialist ideology degrades the society it seeks to serve, while a capitalist ideology ends in consuming its own economic resources. Both fascist and racist ideologies make hateful the nation or race they aim to glorify. A feminist ideology soon exhibits all the disease of the patriarchalism and sexism it abhors. A religious ideology ends in worshipping itself rather than God.

Is it possible that ecology can give rise to its own range of ideologies, perhaps to share many of the characteristics of those listed above? The trenchant comment of a noted ecologist can be taken as a sober warning:

> Vital as 'interconnectedness' may be, it has often been the basis of beliefs ... that become the means for social control and political manipulation. The first half of the twentieth century is in great part the story of brutal movements like National Socialism that fed on a popular anti-rationalism, anti-intellectualism and a personal sense of alienation. They mobilised and homogenised millions into an anti-social form of perverted 'ecologism' based on intuition, 'earth, blood and folk', indeed and 'interconnectedness' that was militaristic rather than freely communitarian. Insulated from any challenge by its anti-intellectualism and mythic nationalism, the Nationalist Socialist

> movement eventually turned much of Europe into a huge cemetery. Yet no one would have believed, given its more naive precedents in the intuitional and mystical credo of a century earlier, that fascist totalitarianism would have gained sustenance from such starry-eyed worldviews.[12]

All ideology is the death of hope and imagination. One all-consuming idea replaces the inexpressible plurality of our real experience. It is precisely here that reflective faith will have a critical role to play. It will have to insist that all intensities of human experience be given their due. As these reflections unfold, I hope that will become clear. At the moment, I merely remark that the only worthwhile response must lie in the use of our best resources: to be very intelligent where intelligence is needed; to be defiantly hopeful where hope is needed; to be passionately loving where love is needed; to be calmly wise where wisdom is needed.

Only in this way can we contribute what we can, and humbly receive what others might give. After all, as Claudel reminds us, sometimes the worst does not happen!

5. The point of faith

Given the ideological perils that seem to threaten from every side, there is a serious question that Christians must face. It affects so much of what follows. The question can be expressed very simply: how does faith fit into all this? The 'this' here means the new ecological and cosmological communities of consciousness which invite religious believers to participate in these novel ranges of concern. I think a helpful preliminary answer can be found in terms of a description of faith given by Bernard Lonergan. Faith, as a knowing born of religious love, is consciousness:

> brought to a fulfillment, as having undergone a conversion, as possessing a basis that may be broadened and deepened and heightened and enriched, but not superseded...[13]

Human consciousness is experienced as a 'coming to' of the self as the subject active in everything we do. We see and hear and touch and smell. It expands to a new level as we imagine, wonder, question. It increases in its specific gravity as we ponder evidence, to get beyond impressions or bright ideas, to the truth and value of things or actions. It is there experienced as putting oneself on the line. It reaches a special integrity in making responsible decisions and in the peace of a good conscience. It blossoms into a kind of ecstatic relationality as we fall in love or give ourselves over to a great cause, the experience of finding oneself by 'losing' it.

In all these levels of expanding awareness, the self we are is experienced as a question: is there ultimate meaning to the meanings that illumine the direction of life? Is there a final sufficient reason, an ultimate confirmation for all the sufficient reasons of the truths to which we hold? Does the mystery that gives rise to the universe ultimately cherish everything we treasure? Am I ultimately in the presence of a 'Thou' in which each of us and our real world comes home?

Faith, in this perspective, is the occurrence, however dramatic or unexpressed, of a sense of Yes to the complex undertow of all such questions. Dag Hammarskjöld, the former Secretary-General of the United Nations, entered the following words in his journal shortly before his untimely death. They give vivid expression to the sense of faith we are describing:

> I don't know Who—or what—put the question. I don't know when it was put. I don't even remember answering. But, at some moment, I did say Yes to Someone—or Something—and from that hour I was certain that existence is meaningful and that, therefore, my life, in self-surrender, had a goal.[14]

Now, while such faith cannot be superseded, while its radical, intimate, universal Yes cannot be replaced by a No, it can be 'broadened and deepened and heightened and enriched'. This is the point at which Christian faith can expand into the new meanings and values of, say, ecology and cosmology. The basic conversion that faith is can be broadened into whole ranges of ecological concern. It can be deepened in its contact with new cosmological understandings of the universe. It can be more collaboratively embodied in the world of human experience and exploration. Such a development, far from undermining religious conversion, enables it to expand into a new context, one more worthy, in fact, of its primary orientation and more attuned to its implicit universality.

Hence, faith continues to seek understanding. In such a search, theology becomes more obviously 'faith making connections', and more fully an 'analogical imagination'.[15] Such a connective or analogical imagination unfolds in its efforts to discover new analogies in which to express the focal meanings of Christianity such as creation, Trinity, incarnation, the death and resurrection of Christ, the eucharist and the final consummation of all things in him. Implicit in such creativity will be new ways of interconnecting such focal meanings not only amongst themselves, but within an ecological and cosmic world of interconnection that so occupies modern consciousness.

This will lead to a more comprehensive expression of Christian hope within an horizon of planetary and cosmic solidarity. 'Life to the full' (Jn 10:10) must include all the dimensions of life as we are coming to know it.

Hence, faith has both a deconstructive and a constructive role. By resisting any kind of a new ideological system in the name of its own universal vision, Christian experience is challenged to embody itself in a freshly comprehensive manner. An ultimate faith, hope and love are not irrelevant to the new holistic visions that are emerging. Far from denying the positive attainments of the new sciences, theology can work to place them in a context of reconciliation, communion and, most importantly, of hope. Even if the light of faith is experienced as no more than a kind of 'background radiation' in many lives of research, art and social commitment, it still communicates its sense of the universe as ultimately friendly.

6. The community of faith

Christian faith is in fact showing considerable vitality with a whole range of resources to offer—theological, moral, spiritual, philosophical and the 'dialogical' experience inscribed in its missionary beginnings, in its medieval synthesis, and in its contemporary commitment.[16] With its sense of creation as a whole, its sense of the genesis of the cosmos as preparation for the genesis of God in the incarnation, with the whole hope it offers, theology is working to a new relevance.

At a juncture when the human race and poisoned earth need all the help they can get, it is as well, then, that those of us who are Christians define ourselves into the great conversation now afoot. In regard to ecological concerns the consideration of the religious dimension has entered into a more productive phase after earlier recriminations.[17]

Though we human beings differ in so many of our visions, values and priorities, there are two things we have in common at this critical moment: first, the immense, laboring fertility of our past which has brought us forth and placed us together in this present moment; and, secondly, the future which in some quite new way will be the product of our present decisions. A 'common era' has been forced upon us by the sheer extent of the concerns we must now share. The imperative, for each of us and for the great classic traditions that have nourished the human enterprise, is to throw our best selves into the mix, so to speak. No one has cornered all the good energies of faith and love, of justice, tenderness and wisdom. And, no matter what way we look at the situation, there is still not enough of such spiritual energies to go around. So, we do the best we can. And that is the spirit in which Christian thinking has to proceed.

No amount of technical organization and political restructuring for the sake of a new society is going to be effective without a large measure of community. If there is no shared experience we will be talking about different things. If there are no shared meanings we won't be able to talk at all. If there are no shared values, we will simply populate the world with antagonists. Hence, a fundamental sense of community is the first requirement for the great creative efforts history is demanding.[18] Unfortunately, this is exactly the area where increasing alienation has ruled: the disaffection of the human from nature, and the violent disaffection of human beings among themselves, with the resulting emergence of huge blocs of competition along political, economic, racial, sexual, cultural and religious lines. Our superstructures have evolved to serve an alienated situation in which the common good is at best a compromise. When economics becomes mere monetarism, when our national GNPs measure everything except what really matters, when politics is locked in competing blocs of self-interest and greed, ecology can easily restrict its aims to uncritical environmentalism: it becomes a matter of merely tidying up the madhouse or landscaping the battlefield. And, of course, huge ungainly bureaucracies keep on growing in a frenzied effort to stabilize a system that is not working.

The most available rhetorics are the liberal one of Late Capitalism with its emphasis on individual rights, and the optimistic one of a 'new world order'—now that the collectivist experiment of the Soviet Empire has expired. Such a discordant situation is hardly hospitable to a deeper sense of differentiated community based on mutual relationships of care and responsibility in regard to the life and future of the planet. There seems little opportunity for speaking realistically about a radical ecological reordering of the world, even if the Earth Summit of June 1992 pointed to the urgency of the situation.

In such a situation the Church has to redefine its identity and role in today's world. The traditional theological definitions spoke of 'perfect societies' such as Church and State, just as modern democracies opted for a radical separation of these two social entities. The shared viewpoint rested on the conviction that both Church and State had legitimate, or at least realistically acceptable, spheres of influence, each with its own end in view. But these accepted divisions and separations are no longer as productive as they once were. We now live in a world in which every group is subject to a common crisis: new forms of planetary cooperation have become necessary to secure a common survival. A new sense of community is now a matter of pressing importance. A profound cultural shift has become imperative.

It is here, I suggest, that the Christian churches must seize the opportunity to offer a unique service at this historical moment with their resources of moral traditions and religious vision. There is no easy way simply to legislate the structures necessary to maintain the quality of the biosphere. Even if such structures might conceivably be legislated, the deep values and meanings relevant to the critical situation cannot be decreed into being. Legislative bodies can change laws, but they cannot change hearts. That can only come about by an inspired new sense of community and solidarity. It is on this level, that of the deep structures of human belonging and shared hope, that the Church can redefine its role.

However, this community-forming ministry of the Church can only operate within larger pursuits of planetary community. Though it is an international institution, with its own long history in the development of Western civilization, the Church is not a world to itself. In times past, it has been the carrier of intellectual and moral values to a significant degree. But now the worlds of intelligence and moral values have expanded to leave the Church not as a teacher, but a learner, not only in regard to the world of science, not only with respect to the morality of human rights, but also in a world of interfaith contact. Hence dialogue has become the key to ecclesial development in the world in which it finds itself.

If the Church is to promote a new sense of human belonging it cannot, of course, ignore the community of science, especially as human intelligence has inspired a new telling of the story of our cosmic and biological origins. Nor, if it is intent on speaking to the human heart, can it ignore the community of moral self-transcendence, of values, of ethical integrity, the company of all people of goodwill, those who recognize that global coexistence has become impossible without a demanding sense of the common good.

Then, too, there is the community of art, that domain of creativity by which our experience is freed and refreshed to perceive the original beauties that the routine concerns of life obscure and distort. For the Church, in its own way, witnesses to transcendent beauty, to the glory that has been revealed, yet too often remains concealed by the obsessive rationalism and moralizing of a religion removed from its fundamental experience.[19]

More proximate to its concerns is the community of religious faith, those who cultivate the transcendent, the holy, the ultimately worthwhile. When self-transcendence flowers to a sense of life within a gracious universe, an energy powerful in its ability to sustain the communities of meaning and moral value is at work. Here the Church makes contact with a community of ultimate connections.

Within these larger collaborations and communities,[20] the Christian community forms itself in the light of its own special experience of God revealed in Christ, of the divine mystery communicating itself to the cosmos, in its ultimate hope that God will be 'all in all' (1 Cor 15:28).

In this larger world of intelligence, morality and religion, the Christian Church has both a redemptive and a constructive role. All forms of community are vulnerable to the baleful influence of 'the seven deadly sins': pride and greed, envy and apathy, violence and dishonesty remain ever apt to frustrate the possibilities of collaboration in the common good. So, by witnessing to its conviction that forgiveness and reconciliation are real possibilities for human existence, no matter how godforsaken the historical situation seems to be, no matter what the extent of the failure and destructiveness present in any past, the Church acts within culture to assist in diagnosing the general plight. In its witness to limitless mercy, the Christian community encourages the demanding integrity required to confess our social sins for what they are, and to renew hope even when healing seems impossible. Without such honesty and hope, any culture is locked in an endless rationalizing in regard to its deepest failures.

When sins are confessed in the hope of forgiveness and reconciliation, a new beginning can be made. Hope enables us to imagine the world otherwise. Our history is not the sum total of our failures. The last word, at least in any Christian statement, is one of grace and limitless mercy.

More positively, the Church has a constructive role in the global and ecological turn in human consciousness. Christian faith lives in a familiarity with the universe as the one creation of God. It surrenders in adoration to that mystery of Love which created the world in order to communicate itself completely to what is other than itself. Such love is the field of life-giving, transforming energy pervading all creation. In those who surrender to it, the energy of God's Spirit inspires a totality, or uni-verse, of all things in Christ, along with an outreach to what is most forgotten, vulnerable—the suffering neighbor in whatever form she or he or it is present. Distinctively, such faith celebrates its sense of the divine presence within creation in the sacramental forms of its worship. There the humble realities of our world become icons of the mystery at work. From beginning to end, and in each conscious moment, the Christian universe is one of ever-original gift, of self-giving relationality, and of ongoing transformation.

Thus, the basic mysteries and essential energies of Christian faith offer deep resources in the building of a sense of global community within a hopeful universe. As such faith learns from the explorations of science and from the moral force of the global responsibilities now stirring in human history, it is enabled to communicate new perceptions of shared meaning and common good. As it discovers the riches of other religious faiths and traditions, Christian faith finds itself in a situation for which its central mysteries have been preparing it, and out of which it can contribute powerfully to a gracious sense of shared existence.

If ecology in its Greek roots signifies 'the meaning of the home', Christian faith can find a new homecoming in entering into such awareness and concerns. The awareness of such a homecoming, not as an alien intruder but as a member of the family, is now a significant emphasis in current theology.[21]

7. The humble self

In many ways, to bring faith and ecology and cosmology together is a move to put our souls back into our bodies. Culturally speaking, we human beings have become a disembodied race, sorry victims of our narrow success. The last few centuries have seen us trying to organize life and our world in a kind of weird objectification of our selves, impervious alike to the wisdom of nature and the wisdom of the spirit. The result is that we now groan under a huge carapace, a heartless impersonal superstructure of economic and social order that has lost its human scale. We have become unselved and denatured.

It is almost as though we have lived through the extinction of humility, understood as our native earthliness. To the humble of mind and heart is given a sense of the human scale: we are not the center of the universe; that lies beyond us. We are not pure spirits, for we are intimately related to the earth itself, in our origins and in our end. As humanity lost humility, it interiorized a grotesquely truncated and distorted sense of the human self.[22] This denatured self-image has tended to turn the human person into an isolated little ghost, more or less haunting the world, and no longer incorporated into the one body of creation. The huge machine of modern technology has not only uprooted us from nature, but cut us off from one another.

'Upward mobility' has been a strange journey. For 'upwardness' has become uprootedness—from the earth itself, from the biosphere in which we exist, from the mysterious ground of existence itself. We have become, to use Walker Percy's phrase, 'lost in the cosmos'.[23]

Certainly, the holistic mentality of our day is contesting the shrinking of the self to such tiny individualistic and disembodied proportions. But the absorption of a general mentality will not be enough. Unless we confront the deep dis-sease of our culture and thoroughly diagnose it, we are exposed to the danger of infecting our new-won ecological concerns with what is still profoundly unhealthy. On the other hand, we may be so vividly aware of the diseased state of affairs, that we start to recoil from what we have become, and give way to disgust. At such an extreme, ecological awareness begins to include within it a kind of self-hatred, a disowning of the human, to end in a generalized misanthropy. Out of such self-disgust, we begin to project onto an increasingly abstract nature what is most unexamined and unredeemed in our own selves.

As faith seeks a new connectedness with the earth and the cosmos itself, it promotes the conversion of self-consciousness to a more authentic standard: the currency of inner capitalism has to be exchanged for real values. The realization that 'we are up to our necks in debt' to the world of nature[24] tempts us to declare bankruptcy; or to leave the country before the debtors find us: but where can we hide? Therein lies the human problem.

Just as despair is fundamentally a failure of imagination, true hope is formed out of the shared active imagination of all those who have humility to recognize this earth as the shared body of our existence. Imagination regains its courage when it is prepared to diagnose the harm caused by the refusal of our earthly status. Our creativities are newly inspired to the degree we are disposed to yield ourselves into a more intimate collaboration with the gracious mystery of Life, however it has been revealed to us.

8. The inclusive language of faith

The reshaping of theology along new lines is an immense and complex task. Fortunately, it has already begun; and in such beginnings, despite the inevitably complex developments that are taking place, two points of radical simplicity are worth stressing: first, the search for a language of universal inclusiveness; and secondly, the extension of the meaning of the greatest of all commandments.

In forging a theological language that will be inclusive of disowned or hitherto unknown dimensions of existence, we need to focus on the fundamental story of creation, as 'the whole story', in a way that invites all the *dramatis personae* to claim that whole story as their own. Admittedly, the issue of inclusive language has been mainly restricted to a feminist criticism of the religious, political and social metaphors that have structured a patriarchal system in the past, and still marginalize women's experience in the present. The Christian story must look to a new telling point at which a positive valuation of the feminine will be assured. The points usually emphasized are the original creation story of Genesis, the Wisdom tradition, Jesus's own inclusive dealings with women, through to Paul's vision of humanity made one in Christ.

But the feminist inclusiveness does not stop here. As most critical feminists assure us, their critique is a prelude to a far vaster form of theological inclusiveness,[25] one that respects the whole world of our relationships, not only to one another and to God, but to the biosphere, the planet, the cosmos itself. We are looking for an inclusive language that speaks into the modern world of meaning the cosmic dimensions of the universalist Christology of the Johannine tradition and of the later Pauline epistles. If all things are made in Christ and through him, if he is the origin, coherence and goal of the entire universe of God's creation (Col 1:16–19), then 'all things' have to be given their due in a genuine Christian vision.

In the present context, the original story now demands a telling in a situation vastly different from that of the biblical authors. The all-inclusiveness of the creation story and the Christian Gospel now have to integrate realities inaccessible to the imagination of former times. Such realities extend from the infinitesimal interactions of tiniest subatomic particles to the outermost exploding star in the furthest known nebula billions of light-years away; they include the helical dance of the DNA and the complex ecologies of the rain forest, the disconcerting world of quantum mechanics, radiation from the Big Bang and impenetrabilities of black holes and dark matter.

Whatever the complexity involved in all this, the challenge to faith can be expressed in terms of simple questions: how does our emerging sense of human integrity and relationship demand a new telling of the whole story? How do we pass from the limitations, preoccupations and even perversions of unwarranted exclusiveness to a more thoroughgoing inclusiveness, one more open to the whole realm of human experience, more worthy of the implicit universality of the faith we profess?

It comes down to this: how do we all belong, including the prepersonal, in the one mystery of the universe? Reflecting on the inclusiveness of that 'we' is a good way of introducing a range of larger questions.

The earlier theology of this century was largely concerned with the mystery of the human person so threatened by the violence and totalitarian ideologies of the time. Hence, in reaction to the philosophical abstractness of past thinking on human nature, the emphasis was laid on the concreteness and irreplaceable uniqueness of each human person. To this degree, it was pre-eminently a theology of the 'I'.

More recently, such a concern has been revealed as all but romantic unless attention is paid to the structures that shape human life: social, economic, political, cultural. You could say theology turned critically to the reality of the transpersonal 'It' of the human world, the socio-political and economic structures that have proved so deeply inimical to the transcendent value of the human person. By concentrating on that 'It', following the lead of the social sciences, theology implicitly invited the 'I' to understand and declare itself in terms of a socially and culturally formed 'we'. It opened reflection to another range of questions: with whom are you in solidarity? Who do you stand with? Who do you speak for?

Here the various versions of the 'option for the poor', of 'solidarity with victims', of a history seen 'from the underside' were worked out. And often they achieved noble expression, as when the blank cheques of such phrases were cashed in terms of heroic resistance and even of martyrs' deaths in the various proponents of Liberation Theology.

Out of that experience and commitment, Christian reflection is entering into the expression of an even larger 'I', of a far vaster and interconnected 'It', of a 'we' realized in an immense inclusive communion.

The key contexts are ecological and cosmic. The 'I' is invited into an awareness of its embodiment in the interconnected, multiform life of the planet itself. The human person is newly perceived as an 'earthling' in the great temporal and spatial genesis of the cosmos. The growing appreciation of such an 'It', the planetary web of life and the cosmic process that has given birth to it, inspires a fresh expansion of the 'we'.

For it is a 'we' born out of humility, responsibility and hope. Out of humility, for it arises from our inescapable dependence on a world of living and non-living things for our existence, nourishment and delight. Out of responsibility, because human freedom at this point of history is radically shaping, for better or worse, the whole world of ecological existence. The fate of the planet and of future generations of living things and persons has come to depend on human decisions. Out of hope, since faith lives from its conviction that we are not alone in the universe: there is an 'Other', creatively, graciously present in every moment.

The great Barbara Ward, writing nearly twenty years ago, captured the simplicity we refer to here:

> When we confront the ethical and natural context of our daily living, are we not brought back to what is absolutely basic in our religious faith? On the one hand, we are faced with the stewardship of this beautiful, subtle, incredibly delicate, fragile planet. On the other, we confront the destiny of our fellow man, our brothers. How can we say that we are followers of Christ if this dual responsibility does not seem to us the essence and heart of our religion?[26]

9. The Great Commandment and a holistic response

Such words take us to the second point of radical simplicity as it is expressed in the twelfth chapter of Mark's Gospel. Asked, in accord with a well-known practice, which is the greatest of the 613 commandments of the Law, Jesus makes the following reply:

> The first is, 'Hear, O Israel: The Lord our God, The Lord is one;

> you shall love the Lord your God with all your heart, and with all your soul, and with all your mind, and with all your strength.' The second is this, 'You shall love your neighbor as yourself.' There is no other commandment greater than these. (Mk 12: 29–31)

The whole cosmic and historical focus of Judaeo-Christian experience is on the supreme value of worshiping the One God. That uniqueness demands an unreserved and total response. It calls forth an unconditional integrity on the part of the believer. Heart and soul, mind and strength evoke all the dimensions of that integrity—a 'holism' that leaves nothing out. It is a matter of responding with *all* that we are, in every dimension of existence. It would probably be forcing the original meaning of this biblical text to make too much of the distinction of the four terms employed to evoke the totality of the desired response. Still, a certain amount of imaginative translation might be permitted in order to sensitize us to the force of this great summons in the present search for a larger integrity and inclusiveness. Hence 'All your heart' (*kardia*) implies something like the prophet's 'heart of flesh' replacing the 'heart of stone'. What is at stake is a new capacity to feel, with a new affectivity and compassion. It is looking to a larger integration of the 'all' into capacities of the human heart. It demands that we begin 'to have a heart' open to the crisis in which our 'beautiful, subtle, incredibly delicate and fragile planet' is involved. Hence, the Great Commandment is meant to be a heartfelt appeal. As such it bears on the conversion of feeling and of turning hitherto unfocused feeling into creative passion for the integrity of God's creation. Without that deepening and broadening of feeling to include all creation, a 'whole heart' would be numbed in its service of God.

Then, 'with all your soul' (*psyche*): this dimension of existence usually implies, in the many strands of biblical, theological and philosophical tradition, the human life-principle. Without such a soul, the body is dead. Because of the eminence of human life in the world of creation, the soul has often been understood as that which distinguishes and separates the human from the inanimate, and from other, lesser forms of creation. And because it is the spiritual aspect of human existence enabling human beings to relate directly to God, it is that which has to be 'saved', plucked out of the chaos of the world at all costs.

Obeying the Great Commandment in the present context demands a revised understanding of the soul, if we are to love the 'giver of life' with a new wholeness. The ecological and cosmic context of our embodiment suggests that the soul is that which vitalizes the human being in an animated communion with all living things. It is the principle of radical connection to all reality. As Greek philosophers and medieval theologians stressed, *anima est quoddamodo omnia*—the soul is in some measure all things, a principle of openness to the universe. Because a human being has a soul, he or she lives not merely in an animal habitat, but in a universe of communication in meaning and in mystery. The soul is the root of our capacities to celebrate and transform the world. Hence, loving God with your 'whole soul' is, in a profound sense, offering to God life conscious of itself as a gift, life exulting in the mystery of communion, life in contact with a universe. It is not a spiritual escape route from an indifferent or sinful creation into a divine realm, but our human capacity to accompany creation in its journey to God, with the energies of love, thanksgiving and wisdom.

In a similar vein, loving God with 'all your mind' (*dianoia*) means loving God with a mind made whole by reflecting on the Whole. It is a mind dedicated to an inclusive understanding, as it explores and celebrates the wondrous diversity and beauty of all creation. It is a wholeness won by dint of disciplined openness to the new, by patient learning, through loving familiarity with everything as a manifestation of God's creative presence; and in humility before the mystery of it all. It is mind turned contemplative as it beholds creation as the temple of God and as the manifestation of his glory. More practically, for Christian faith, it is mind ready to relearn its basic meanings of God, incarnation, hope and moral life in the light of the new sciences of wholeness.

Finally 'with all your strength' (*'ischus*) comes to have the meaning of using the best energies of the heart, the soul, the mind in the service of the Creator; and in collaborating with the divine in our service of creation. The strength, the energy, the enterprise that characterize human history have all too often been intent on self-promotion, ending in the domination of others. Now, the integrity of human 'strength' is called to manifest itself in other-regarding service, in making connections, in promoting communion, in hallowing rather than desecrating the earth as the shared body of our coexistence. Loving God with the resources of such strength means offering to the Creator a personal contribution for the sake of the 'all', the whole, the Reign of God in which all creation will come into its own.

The link that I have been making between the integrity of our self-offering to God with the inclusiveness of our relationship to all creation is supported in the way Jesus links this great commandment to be wholly centered on God with a second commandment, 'You shall love your neighbor as yourself'. The implication is that we can only come to the one God in the company of the 'neighbor'. Relationship to the other is inherent in relationship to oneself. One's 'self' in the world of God's creation is essentially relational, a self with and for others. Loving God means living in a limitless open circle of often disconcerting inclusion. We cannot imagine we love God because we love nothing in particular. This love of the created other, our neighbor, has now to develop into a new sense of communion and companionship with the whole of this creation. It has to mean not only care for the suffering human neighbor, but also care for all forms of life, and for all the life-nourishing elements and conditions that have so often been taken for granted in the ungracious culture of our times. Needless to say, neglect or exploitation of these larger, prepersonal dimensions of the 'other' has been to the detriment of our human neighbor, present and future. We can no longer imagine we love our neighbor without having a care for the neighborhood.

In the light of this contextual appropriation of the great classical commandments of Judaeo-Christian tradition, God is not just the ultimate focus of an individual soul, but the space of mystery in which all the cosmic and ecological 'we' and 'it' comes to be. The whole of the human heart and soul and mind and strength—all our capacities to relate, to pray, to imagine and to think, to make and to create—are invited to expand into infinities of Love. That is the mystery from which the universe emerges, the atmosphere it breathes, the source of its ultimate transformation. Perhaps we can begin to read S. T. Coleridge's words at the end of *The Rime of the Ancient Mariner* in a new light:

> He prayeth best who lovest best
> All things both great and small;
> For the dear God that loveth us,
> He made and loveth all.

10. From simplicity to complexity

We turn now from the simplicity of a new inclusive language and of a more generous universality in our love of God and neighbor, to the complexity involved in articulating and implementing such wholeness. For the development of a Christian theology adequate to the task, an immense, long-term collaboration is required.

In the meantime, the present situation is extremely complex and even chaotic—in the (now) good and bad senses of the word. The early phases of the current situation were marked by bitter recrimination on all sides. You could sense an extreme irritability which is perhaps characteristic of cultures that have become so complex that any common meanings and values are increasingly elusive. The variety of complaints is well known. For instance, some are determined to dismiss the Judaeo-Christian tradition as inimical to ecological awareness and scientific versions of the cosmos. Religious faith is understood by them as necessarily demeaning the natural in its concern for an other-worldly transcendence, as though the world is simply the prelude to a post-mortem spiritual fulfilment. Others reject science itself as a kind of learned stupidity that has lost itself in the parts of reality it explores, to the forgetfulness of the whole. Bacon, Newton and Descartes are the main villains in this scenario of recrimination. Others reject technology as alien artificiality interfering with pristine nature; while others reject the West, or the North, as imposing oppressive economic structuring on the East or the South. Others single out Capitalism or industrialism or modernity or post-modernity as the culprit. Then, too, powerful feminist voices among the ecologically minded proclaim the culprits as patriarchy, hierarchy, chauvinism—male domination in all its forms.

The most disconcerting rejection is that of humanity itself as some kind of selfish pest infesting the planet and perverting the happy ordering of the natural world.

Of course, as all these huge groups are singled out for their destructive influence in the present situation, they have defended themselves of criminality, and found 'Greenies' at best romantic dreamers, at worst New Age fascists willing to sacrifice everything to their aims.

There is no point in pretending that such antagonisms were not, and are not, partially justified. Yet they are not the whole story, and cannot be if there is to be any possibility of more enlightened collaboration. Hence we must hope that the situation can rapidly move into a constructive, collaborative phase as Christians and other religious believers, as scientists and industrialists of the North and the West begin to own the common crisis; as critical feminists continue to explore the full extent of our common problems; and as ecological activists realize that they are not alone in their concerns.

Christians have to repent of their sins; and it might be no small service to give the lead here. Each one of 'the seven deadly sins' has an anti-ecological connotation to it: pride (the rejection of the humility of the human scale); covetousness (defining oneself in terms of having, rather than being); lust (the denial of the sacredness of life and relationship); anger (the extreme intolerance that sees all diversity as a threat); gluttony (the destructive consumption of precious common resources); envy (a self-absorption that permits no gratitude, or joy in the diversity of gifts); and sloth (expecting nature and life to give us a free ride)! There can be no real collaboration without acknowledging failure, powerlessness; without asking forgiveness for falling short of the great commandments of love, and for failing to live out the whole logic of the incarnation of God amongst us. Still, I think Christian life—in the essential energies of its faith, hope and love—has more to offer than the seven deadly sins of Christian failures. Grace keeps on being grace; the healing and hope it offers can be a beneficent influence in the great concerns of the moment.

Moreover, we should bear in mind that the last two hundred years, so destructive of the biosphere, have been notably unfriendly to Christianity as well.

Scapegoating either Christianity or science, or technology or humanism, or men, or the North or the West, while it might be understandable given the widespread panic and desperation many feel, tends to obscure both the gravity of the problem and the possibility of any solution. Nothing is quite as simple as our first apprehensions of a problem incline us to think, especially when we realize, despite the advances of scientific knowledge and ethical commitment, how little we really know. Focusing on single issues does not readily lead to inclusive thinking.

Every attempt to enter into the ecological conversation faces us, whatever our point of view, with the deepest kind of philosophical questions. Amongst these, the most important turn on the meaning of nature in general and of human nature in particular. That, in turn, leads to the necessity of exploring the relationship between nature and human history, and between both these and the cosmos as a whole. If such explorations bring us back to ourselves as the shapers of the future in some decisive way, we soon find ourselves pondering on the criteria for ethical decision and the role of human intelligence as a directive force in the evolutionary process: how are we all caught up in a common cosmic purpose? Huge questions move and shift below the surface of any conversation.

In the midst of such complexity, the wisest stance is to begin where we are and to see what we can do. As Merry, the Hobbit in Tolkien's *Lord of the Rings*, advised, 'It is best to love what you are suited to love, I suppose: you must start some place, and have some roots, and the soil of the Shire is deep.'[27]

'It is best to love what you are suited to love'. What is at stake at every point is a more expansive and inclusive love. The love of our neighbor has now to include so much else as we awaken to the mysterious universe of our coexistence, and the varied forms of life it has brought forth. The love of our God, too, must expand in gratitude and in a larger, more generous collaboration with the divine will that has brought us into existence, that has brought us together on this earth, that has made each of us a necessary presence in the destiny of the other.

'You must start some place'. In the long, demanding movement into a new phase of history, it is to be expected we will have different starting points. Democracy and pluralism are resources, not limitations. What matters is that each of us seeks to rejoin the human race, enter the stream of life, from where we are, in a new celebration of life's manifold mystery. History invites, whatever our distinctive paths and pace, that we move together in the one direction. It counsels, too, a greater reserve in excommunicating one another from the planetary community we now are. For we have at least two things in common: the common body of our earth, and the future that is coming upon us. Like it or not, we are fellow travellers, and we have to make the best of it.

The realization is dawning that we can no longer be passengers in the ark more or less discomforted by the odd variety of other living things that bark or howl or sing or hiss or roar and growl in other compartments of the boat. All share the same crisis. It is up to the human family to determine the direction and reach landfall before it is too late. Without that shared sense of direction, every wind is contrary and we remain adrift in a world of growing troubles.

My own starting point, shared, one way or another, by at least a quarter of the world's population, is Christian faith. 'You must start some place, and have some roots.' The roots that I will be emphasizing here are the perennial roots of Christian faith: the self-giving communion that is the trinitarian God; the universe as God's creation; the incarnation as the genesis of God within that creation; the sacramental nature of all reality; human consciousness become conscience in responsibility to the Other and the Whole. These are roots indeed; and in these times of crisis, it would be a failure of cosmic proportions for Christians to let these roots wither.

'And the soil of the Shire is deep': Merry might forgive us for extending his much loved 'shire' to mean the whole of our beautiful planet. Here, in this place, we are feeling the need to earth ourselves again, to recover our identity not as Hobbits, but as earthlings, humbly and lovingly connected with all life in the 'meaning of our home' (the *logos* of the *oikos*). The soil, the *humus,* is deep; the deeper we go into the soil of the life and existence we share, the greater communion we will enjoy in moving toward a more gracious, a more 'humble' existence together. For Christians that depth is especially revealed in the entry of God into our world, and indeed, into all the dimensions of our existence.

11. The vocabulary of roots and connections

The scope of a repentant and reconciling Christian holism can be helpfully traced by dwelling on the roots of three key words: ecology, religion and catholicity.

First, 'ecology'. This word has been in use little more than a century after being coined by the German zoologist Ernst Häckel. Its Greek roots imply 'the meaning of the home', as we have just mention-ed.[28] Consequently, it came to refer to the study of the complex totality of conditions necessary for the survival of particular living organisms. By stressing the complexity of relationships characteristic of a given organism, it not only emphasizes the importance of such 'coexistence', but also raises the question of the extent to which living things are fundamentally living relationships. The exploration of such interconnectedness throws light on how living things, whatever their species, are truly 'at home' in earth. As a new science of wholeness, it explores the 'home', the *oikos*, as the matrix of all the relationships of living, where each living thing is at home and has a livelihood.

In highlighting the interconnectedness of everything in the one web of planetary life, ecology soberly warns us that 'everything goes somewhere' not only to nourish but also to pollute a larger world. Thus, ecology is concerned to explore nature 'in the round', so to speak, in all the intricate, delicate interactions that characterize life on this planet. It is the study of our planet as a 'community of communities' where a species is not just an isolated specimen, but a facet of living, interconnected totality.[29]

The second key word is 'religion'. This time the Latin roots are instructive: *religare* (to bind together again) or *re–eligere* (to renew one's choice). A fresh comprehensiveness of faith and action is implied. The religion of this time must aim to tie our experience together in a greater wholeness, and to choose the path of wholeness for a shared healing and a common health. A deeper, more tender relationship of deliberate bonding to the earth and the universal process will tend to make us see life, not as a bundle of problems to be solved, but first of all as a gift, a shared connectedness within a life-giving mystery. The problem in recent centuries has been that our religious sense has not notably linked us back to the earth, nor linked us with the whole communion of living things. This is an odd attenuation of Christian experience. For, with its accent on creation, the incarnation of the Word, the resurrection of the body, the sacramental character of the divine presence, Christian faith is, in so many ways, the most earthy and material of all religions.

Our third word is 'catholic', from the Greek *kata holou*, literally, 'universal', 'all-embracing', 'in accord with the whole'. It has, of course, its original historical meaning in the self-description of the 'Catholic Church', in its institutional intent to welcome the whole—the totality of God's revelation and the totality of Christian response in all the variety of cultures and languages in which it occurs. But whatever our particular Christian traditions—Catholic, Orthodox, Anglican, Protestant—'catholicity' is generally accepted as a mark of the authenticity of faith. It remains, therefore, a mark of fundamental Christian concern; but now as awaiting a larger application in an ecological and cosmic frame of reference. Such 'catholicity' evokes a more expansive way of indwelling creation as participants in the totality of the mystery revealed there, in the one mystery of Christ 'in whom all things hold together' (Col 1:17). Here Simone Weil's question is pertinent: 'How can Christianity call itself Catholic if the universe is left out?'[30]

Such a cursory reference to the classical derivations of these key words suggests a far more serious search for roots. How might the sense of the living planetary whole extend the wholeness that Catholic faith pretends to celebrate? How is the whole marvelous web of life on this planet to be integrated into the Catholic sense of 'grace healing, perfecting and elevating nature', in a way that makes connections with the Catholic doctrines of creation, Trinity, incarnation, sacrament, and natural law? How, in short, in the face of the eco-catastrophe that threatens planet earth today, might a Catholic sense of the universe welcome and promote a more profound ecological commitment?

12. Conclusion to the first circle of connections

One can hardly doubt that some enormous change is called for in human culture, in our particular lifestyles, in the expression and practice of our faith. There is an anxiety inherent in the 'turning point' we have been considering. Will it really become a turnabout, a conversion of the religious, moral, intellectual and spiritual dimensions that are necessary?

The concluding lines of E. F. Schumacher's *Guide to the Perplexed* seem to me to offer salutary advice:

> Can we rely on it that a 'turning around' will be accomplished by enough people quickly enough to save the modern world? This question is often asked, but no matter what the answer, it will mislead. The answer 'Yes' would lead to complacency, the answer 'No' to despair. It is desirable to leave these perplexities behind and to get down to work.[31]

In that spirit, we will continue with this search for connections, as a tiny ingredient in the great conversation on the meaning of life and of our place in it. Each of us has to enter the conversation from where we are. The more we own that standpoint with humility (conscious of its limitation), with hope (we are not alone in the universe), with compassion (we are all involved together in the great drama of life), with imagination (we are stirring to the intimations of a new dream), the more genuinely ecological the conversation will be. It is a matter of meaning our lives as an ever-expanding, open circle.

We must now enter into the second circle of connections.

1. See N. M. Wildiers, *The Theologian and his Universe: Theology and Cosmology from the Middle Ages to the Present,* Seabury, New York, 1982. The 'his' in the title of this excellent book unwittingly substantiates the author's thesis!
2. A significant book here is Paul H. Santmire, *The Travail of Nature. The Ambiguous Ecological Promise of Christian Theology*, Fortress Press, Philadelphia, 1985.
3. See Vatican I, *Constitution on Divine Revelation*: human intelligence illumined by faith reaches some understanding of its mysteries 'by analogy with truths it knows naturally, and also from the interconnection of the mysteries with one another and in reference to the ultimate human destiny'.
4. For a good expression of the challenge, see John Honner, 'A New Ontology: Incarnation, Eucharist, Resurrection, and Physics', *Pacifica* 4/1, February 1991, pp. 15–50.
5. Of special value here is Juan Luis Segundo's monumental work, *Jesus of Nazareth Yesterday and Today, I-V,* trans. John Drury, Orbis, Maryknoll, 1984–88. For an impressive overview, see Frances Stefano, 'The Evolutionary Categories of Juan Luis Segundo's Theology of Grace', in *Horizons* 19/1, Spring, 1992, pp. 7–30.
6. 'Echoes of the Big Bang', *Time Magazine*, May 4, 1992, pp. 50f.
7. See for example, Ian Bradley, *God is Green: Christianity and the Environment*, Darton, Longman and Todd, London, 1990; and Rupert Sheldrake, *The Rebirth of Nature: The Greening of Science and God*, Bantam Books, New York, 1992.
8. As regards the Earth Summit conference itself, it is true that the 8,000 journalists present did not give it a good press. Still, this unique meeting did bring together 40,000 participants from 178 countries, including 116 heads of state. Further, the Rio Declaration of 27 principles and the 860 page *Agenda 21*, and two legally binding conventions on biodiversity and climate change remain as guidelines for the next decade. A Sustainable Development Commission has been set up to monitor the implementation of new ecological programs. See Keith O'Neil, 'The Road from Rio', *Justice Trends* 66, September 1992, p. 3.
9. *De Anima*, 111, 8.
10. Niels Bohr, for instance: 'If anybody says he can think about quantum physics *without* getting giddy, that only shows he has not understood the first thing about them.' For this and similar remarks by Richard Feynman, Albert Einstein and Wolfgang Pauli, see Christopher F. Mooney, 'Theology and Science: A New Commitment to Dialogue', *Theological Studies* 52/2, June 1992, pp. 298; 405.
11. On the subject of the imagination and hope see the modern classic, William Lynch, S.J., *Images of Hope: Hope as the Healer of the Imagination*, University of Notre Dame Press, Notre Dame, Indiana, 1974.
12. Murray Bookchin, *Philosophy of Social Ecology: Essays in Dialectical Naturalism*, Black Rose, Montreal, 1990, p. 10.
13. Bernard Lonergan, *Method in Theology*, Darton, Longman and Todd, London, 1971, pp. 107; 115.
14. Dag Hammarskjöld, *Markings*, trans., W. H. Auden and Leif Sjöberg, Faber and Faber, London, 1964, p. 169.
15. For the full import of such a phrase, see David Tracy, *The Analogical Imagination: Christian Theology and the Culture of Pluralism.* Crossroad, New York, 1981.
16. The list of outstanding contributors would be very long, and my own assessments will be clear from the documentation I offer in the various sections. These will range all the way from the great 'paradigm shifters' such as Teilhard de

Chardin, Bernard Lonergan and Karl Rahner to such gifted communicators as Thomas Berry, David Toolan, Sean McDonagh, and Denis Edwards, right through to vigorous popularizers such as Matthew Fox. I acknowledge, however, a special debt to those who bring the complexities of science and the sense of faith together in their distinctive ways—writers such as Charles Birch, John Honner, John Polkinghorne, Arthur Peacocke, Stanley Jaki, Ian Barbour and Christopher Mooney.

17. J. Ronald Engel, 'The Ethics of Sustainable Development', in *Ethics of Environment and Development*, I. Robert Engel and Joan Gibb Engel (eds), University of Arizona Press, Tucson, 1989, pp. 13ff. Also James A. Nash, *Loving Nature: Ecological Integrity and Christian Responsibility*, Abingdon Press, Nashville, 1991, is especially valuable in the section 'The Ecological Complaint against Christianity', pp. 68–91.
18. In reference to the deep structures conditioning management theory, see Jeremiah J. Sullivan, 'Human Nature, Organizations and Management Theory', *Academy of Management Review* 11/3, 1986, pp. 534–49.
19. Here the towering figure in the theology of beauty is Hans Urs von Balthasar.
20. See I. Robert Engel and Joan Gibb Engel (eds), *Ethics of Environment and Development*, University of Arizona Press, Tucson, 1989, for an excellent example of the current dialogue.
21. For a valuable survey of authoritative statements and theological resources, see Denis Edwards, 'The Integrity of Creation: Catholic Social Teaching for an Ecological Age', *Pacifica* 5, 1992, pp. 182–203.
22. For a challenging treatment of this theme, see William Barrett, *Death of the Soul: From Descartes to the Computer*, Doubleday, New York, 1986.
23. Walker, Percy, *Lost in the Cosmos: The Last Self Help Book*, Arena, London, 1983.
24. David Toolan, '"Nature is a Heraclitean Fire". Reflections on Cosmology in an Ecological Age', in *Studies in the Spirituality of the Jesuits* 23/5, November 1991 [whole issue].
25. See Anne Primavesi, *From Apocalypse to Genesis: Ecology, Feminism and Christianity*, Fortress Press, Minneapolis, 1991.
26. Barbara Ward, 'Justice in a Human Environment', in *IDOC International* 53, May 1973, p. 36.
27. J. R. R. Tolkien, *Lord of the Rings*, Vol. 3, Ballantine, New York, 1965, p. 179
28. Douglas M. Meeks, *God the Economist: The Doctrine of God and the Political Economy*, Fortress Press, Minneapolis, 1989, pp. 33ff.
29. For an interesting attempt to set new perceptions into a larger tradition of philosophy, see Laura Landen, 'A Thomistic Look at the Gaia Hypothesis: How New Is This New Look At Life?', in *The Thomist* 56/1, January 1992, pp. 1–18.
30. Simone Weil, *Waiting for God*, Fontana, London, 1959, p. 116.
31. E. F. Schumacher, *A Guide for the Perplexed*, Harper & Row, New York, 1977, pp. 139–40.

SECTION II

A Second Circle of Connections: Contexts

I would now like to extend in the following spiral of connections, some of the themes and questions already introduced. The aim here is to work to an even more critical, inclusive sense of reality and to approximate to a greater holism, a more complete catholicity, in the outreach of reflective faith. First of all, we stress the new inclusive quality of the basic cosmic story we share.

1. The story

For some time now, it has been a lament of even the most critical thinkers[1] that the increasing pluralism of our culture no longer permits a meeting of minds and hearts on the discussion of even the deepest moral issues. The reason for such polarization is the lack of any shared story. Without such an inclusive narrative, there can be no common frame of reference, no shared sense of identity, no principles from which to resolve the urgent issues of the day. When each conflicting group has its own exclusive account of the way things are, the other can be offered only the dubious role of being the adversary in someone else's story.

And yet, a new comprehensive story is beginning to be told. In a way, it is the scientific story of our origins and common belonging.[2] In the precious objectivity that genuine science offers us, it is a story that can be owned by every group and individual that prizes truth as a fundamental value. How much it conflicts with universalist religious narratives, how the story of creation relates to this new scientific story, are questions that will occupy us in later phases of exploration. But, at the moment, it might be helpful to give a couple of imaginative versions of the scientific story that is emerging as the greatest resource for a continuing conversation on how we belong, and must belong, together. Attention to the significance of the four and a half billion year story of our planet, of the fifteen billion year story of the emergence of the cosmos in all its differentiation, growing consciousness and connectedness, must lead to a point beyond the present situation. So much time has gone into the making of the present that it would seem to be the deepest disloyalty to our history to leave the last word to mutual recrimination. Surely, such a long unfolding of life is not meant to climax in either the frenzied consumerism of our culture or leave us at the brink of self-extinction in an increasingly perilous present. It has to mean more than *this*.

What, then, is this inclusive, hopeful story? It can be told most vividly when the objective statements of science are embodied in the flesh and feeling of imagination. It is imagination that gives body and momentum to meaning: it takes the necessary abstractions of theory out of their linear sequence; it drags the objective data from the computer screen to reform them in the taste, the drive, the passion of incarnate existence. The result might be something like the following example.

It has been generally estimated that our planet is about 4,600 million years old. To bring such a length of time into human imagination, it is helpful for each of us to assume an imaginary role in the process—to be shamelessly angelic for a moment, so as to think of oneself as a lifelong companion to the earth as we have come to know it. It would mean thinking of yourself as about forty-six years old, with each year representing, in fact, a hundred million years.[3]

In this artfully imagined world, you will have to wait till you are forty-two before you thrill to the sight of a flower and smell the fragrance of a bloom. Then, four months into your present forty-seventh year, you will have the special companionship of the mammals: you will see the whale blowing off the coast; the monkeys swinging in the trees; the great cats stalking through the jungles, as you begin to wonder whether you could ever tame a dog or ride a horse. But then, just four hours before midnight on the last day of your forty–seventh year, another more intimate form of companionship will have been offered you when you began to respond to the smile of a human face.

Then, in the next three hours leading up to your forty-seventh birthday, after various attempts at conversation, you will have worked out with your human companions how to plant a tree and tend a crop. In that last hour, you will have exchanged many skills—in building and trading; in recording and writing; in making art and, more ominously, in making war. But only this last minute, the whistle of the great factories will be heard and their smoke begins to darken your town. Then, in the last few seconds, you will find yourself coughing and sneezing because of the poisons in the air. You will notice that the water has a strange taste, that some trees are dying, and that many of your old companions among the birds and the fish, the flowers and trees exist no more, as you thread your way through the rubbish dumps of the great cities, and hear rumors of wars that could destroy the planet itself.

And the dark question stirs: what—of all the beauty and variety of such growth and companionship—will be there in the morning, to celebrate your forty-seventh birthday, should either the heat or the freezing of former times return? Like Dante, you find yourself midway through life, lost in a dark forest. You have wandered far from paradise.

Speaking of the last two hundred years, Thomas Berry, one of the geological story-tellers of our day, makes an incisive remark:

> During this period the human mind lived in the narrowest bonds it has ever experienced. The vast mythic, visionary, symbolic world with its all-pervasive numinous qualities was lost. Because of this loss, humanity made its terrifying assault upon the earth with an irrationality that is stunning in enormity while we were being assured that this was the way to a better, more humane, more reasonable world.[4]

Such a judgment shocks the imagination into further efforts.[5] In the present sadness, your angelic trance will look for consolation in still older cosmic memories. It will stir with recollections reaching back to the ten billion years or so before the four and a half billion year history of the earth. But now, since angels have special capacities, you concentrate your long-term memory of this fifteen billion years into the span of one fantastic, cosmic year. At the beginning, you recall the great shining, the flame and blazing forth of what later generations, rather prosaically, would come to call the 'Big Bang': the first incredibly compact point of explosive energy, the first moment of a cosmic birthday of what was to unfold, a universal January 1, at the beginning of this single, all-inclusive year. Even reduced to such a span, it is a long time back, since you know that no human word was spoken, no human face smiled or wept until ten seconds ago, the closing moments before a second year would begin.

Your dream goes through the months, stupefied by the unfurling grandeur of a great fireball, expanding and condensing in swirling, molten masses ... until, at the beginning of a mystical May, the great wheel of the Milky Way prompts you to recognize something you know. Then, on September 9, our little star, the sun, begins to shine distinctly in the ocean of radiant explosion. One week later, a molten droplet in the midst of all this is cast off; and our tiny planet has been born.

As it cools, in the course of a week of so, our earth begins to show its promise: something is alive, moving, growing, aware ... September 25!

Weeks pass, as the hesitant little movement begins to pulse more strongly. Then, in early October, these tiny living things begin to unite in ways that were as strange then as they are now: sex enters the life-force, and the opaque thing we call death begins to establish its inescapable limits. But it is another month before the air becomes fit to breathe, and larger complex living forms begin to grow and crawl in the open day: December 1.

Two weeks into December, the first worms are crawling; a few days later, the first fish move in the waters; by December 20 the first plants spring up; by December 21 the first insects buzz and swarm; and then come the animals of a kind. But not till December 23 do the first trees grow and reptiles slither by the lakeshore; dinosaurs begin to leave their footprints on December 24, while the warm blood in the first mammals begins to flow on December 26. On the next day, the first birds begin to fly. December 28 awakens to the color and fragrance of the first flowers as they begin to bud and bloom.

The great dinosaurs suddenly die out; and mammals start to give milk to their young. These flourish, as their brains become more intricate and responsive in the stimulus of the next day. Then on December 31, the first of human kind begin to play and speak and love in ways that we would recognize; and the rest, in a quite literal sense, is history.

To awaken from such a dream of the earth in terms of a middle-aged human life, or of the history of the cosmos compressed into the span of a year, takes one close to the point of prayer. In the above imaginary span of things, Christ appears only an instant ago: the earth opened to bud forth its saviour.[6] True, the baleful effects of human influence have entered the very geology and biology of the planet, as the air, water, soil, and varied generativity of the earth have been deeply, if not irreversibly, infected. On the other hand, the human race has scarcely begun to appreciate the new feature of the universe that is the focal point of Christian witness: the mysterious Ground out of which the universe has emerged has entered, at this critical last moment, into the wonder and anguish of our world, to be itself a participant in the cosmic drama.

We are still at the great turning point in our history, the beginning of our second cosmic year when the Word as the divine self-expression has been uttered into the struggle and groaning of the universal process, so to offer a new hope of reconciled existence. The Light shines in the darkness, and the darkness cannot overwhelm it (Jn 1:5). Such hope is the telling point at which to hear the struggling life-story of our planet, and the all-inclusive story of the cosmos itself. It leaves us with a question: how can the energies of Christian faith, hope and love contribute to the getting of wisdom?

2. Homo sapiens: fact or hope?

It would seem that our self-designation as homo sapiens may have been a trifle premature. Instead of self-congratulation, it has come to signify a hope that we will have the wisdom to face the crisis our misdirected energies have largely brought about.

Energy there certainly was. The past four hundred years were times of remarkable scientific discovery, geographical exploration and economic innovation and development. Our planet has been gradually drawn into a single system of technical and material interdependence. But the dimensions of such convergence have also broadened and deepened the scale of the conflicts that were hitherto localized or ignored. The global scale of both the developments and of the rising expectations—intensified by national, regional and ideological conflicts—have inexorably increased the strain upon the planet's depletable resources and upon the delicate biological mechanisms that have sustained life on earth. Human history is being forced to pause to take fresh bearings, and to redefine the meaning of progress itself.

Given the scale and intensity of the problems that confront us, wisdom has become not the presumed endowment of *homo sapiens*, but a hope for some kind of new enlightenment. We have begun to feel the need for both a new sensibility toward the wonder and fragility of the biosphere, and of a new solidarity in global destiny. The 'coexistence' of mutually antagonistic ideological blocs of the Cold War era now has to give place to a more positive, collaborative form of coexistence in terms of the great cosmic story that shapes our various histories into one destiny.

In the Western philosophical and theological tradition, the quality of wisdom, of *sapientia*, was understood as the ability to judge and order reality from the highest standpoint.[7] It meant having a certain taste (*sapere* means 'to taste') for the deepest realities of life. This wise taste for reality was manifested in a kind of intuitive familiarity with the whole of nature, resulting in what was termed 'connaturality'.[8] It was understood as a sense of wholeness and depth that exceeded the multifarious, analytical calculations of human ingenuity and reason. Forms of such wisdom could be either a supreme human attainment as in a philosophical synthesis or the highest gift of the Holy Spirit when mystical faith tastes the divine mystery. The knowledge characteristic of wisdom was eminently a sense of reality born out of a loving familiarity with, and a deep immersion in, the whole mystery of life. This classical notion of wisdom easily translates into the terms of our contemporary need of a fresh sense of the totality of life, and of a more profound, relational participation in the wholeness of existence. A larger, fresh sense of proportion lies at the heart of any 'new paradigm' for human experience as a whole and for science in particular.

3. The new paradigm

The elusiveness of such wisdom is exactly the problem the modern era has begun to face—the inability to grasp the wholeness of things and our own human position within it. 'Know-how' has to find a larger wisdom, just as the 'know-it-all' needs a new humility before the vastness and complexity of reality in all its references and connections. The developing 'holistic' view of nature and 'holographic' sense of the universe are alike repelled by what is apprehended as the prevailing flattened, fragmented, mechanistic, instrumental relationship of science and technology to the realities which they pretend either to explore or transform.

The notion of a 'new paradigm' is frequently explored in contemporary dialogue and conversation. A good example of convergences in such a method is found in a recent book recording the exchanges between an influential scientist and two Benedictine monks.[9] Evidently, the word 'paradigm' has come to carry a rich variety of meanings, values, associations and feelings. To speak generally, it connotes the fundamental symbolic sense of reality which structures the methods and priorities of exploration.

The necessity for a 'new' paradigm occurs when the former one no longer works. The map has become too inaccurate, or incomplete or roundabout. So we begin to look for a new point of entry into the unknown, one more open to the varied possibilities of the journey, one more respectful of the terrain we need to cover.

Why the old paradigm no longer works, why a new one is so urgently needed, can, I suppose, be explained in a fairly simple pragmatic fashion. After all, what doesn't work, doesn't work. This is embarrassingly obvious in geography when early Australian explorers, expecting to find an inland sea, found, instead, a desert. Again it is starkly obvious in the various sciences when it begins to be remarked that space is not filled with ether, that subatomic realities are not small billiard balls, that the sun does not circle the earth, and that the world did not begin five and a half thousand years ago. On the other hand, when science—and the human experience out of which it grows—wakens to its current holistic proportions, the need for a new paradigm is more difficult to either express or explain. The disconcerting element in the new holism is the presence of the human self of the explorer. The journey 'outward' into reality needs a more firm connection to a journey 'inward'—into those mysterious dimensions of human consciousness which are capable of endless differentiation, as in the objectivity of the scientist, the creativity of the artist, the unutterable experience of the mystic, and in the everyday communication and action that make up the ordinary human world.[10]

We can all get better at what we are doing, to find that the old ways no longer work and that some new approach is desirable. But the new paradigm is more than that. What is more, we can all be more deeply converted at religious, intellectual, moral levels of our existence. For example, we might begin to experience God more fully; or perceive evidence we had long overlooked; just as conscience might stir to ranges of values, say, in social justice or ecology, that had not previously troubled our routine ways. But, here too, the new paradigm is, I think, more than any of these.

Then there are all the variety of differentiations of consciousness which make human conversation both difficult and enriching. Because the reality in which we are immersed is so multiform, and because our capacities are limited in taking in such a totality, human consciousness knows many differentiations or mentalities. This is borne out in the languages used in describing the various 'real worlds' we inhabit.

The language of the journalist specializes in communication within the world of the daily concerns of the population. The language of scientific theory, unhampered by the responsibilities of communicating with the laity, is designed to serve the objective exploration of matters so austerely specialized that years of training are required to make any sense of it at all. For its part, human psychology elaborates a language to express reality 'from the inside out' as it were, in terms of the consciousness, the images of self, the feelings and concept-formation that enable human beings to speak or learn or love or create in the first place. As a further complication, artists insist on belonging to none of these real worlds in their concern to refresh our perceptions of the sheer originality, the beauty, of experience, before it can be explained, or used, or explained away. Then mystics, if they speak at all, attempt to give some expression to the intimate presence of ultimate mystery which they have tasted and felt as the origin, ground and goal of the universe itself.

While the new paradigm approximates to a newly differentiated mentality or consciousness, it seems to me to go further. It goes further because the new holistic paradigm includes all of the above. It is certainly a new learning along the lines of the skills and methods with which we might be familiar. Both the data and the methods of every science have been immeasurably extended and enriched. Certainly, too, it has elements of the beginning or deepening of conversion on religious, intellectual or moral levels. For the new shift inspires fresh considerations of religious ultimacy and destiny. It invites new ways of learning and understanding. It reveals new ranges of values in its ecological concerns and its cosmic awe. Finally, it is akin to more easily identifiable differentiations of consciousness. You might say it aims to cultivate a consciousness that brings together all the differentiations of consciousness into the human conversation, and to reveal their roots in each individual self. In promoting a consciousness of the all, and of the interconnectedness of all reality, it is preeminently an awareness of relationships, especially of those relationships that are newly known, or long ignored: with the earth itself, with the biosphere of this planet, and within the emergent process of the cosmos itself.

Even this summary indication of the ingredients of the new paradigm makes it appear less of an arbitrary option or an inspired hunch. What is more, by expressing it in the terms I have employed, I hope to have made the search for a new paradigm more open to the religious and Christian dimensions that I am here commending. Most of the terms I have employed (conversion, the structure and differentiations of consciousness, and so forth) enjoy a widespread respectability through the writings of Bernard Lonergan, above all his *Method in Theology*. His method, understood as 'a framework of collaborative creativity',[11] is founded on the phenomenon of consciousness, and on the dynamics, differentiations and conversions it manifests. Such a method is inherently open and flexible, and proportionately applicable to all fields of learning. To my mind, such theological developments are not only hospitable to the new holistic paradigm, but can easily enhance it. Through the work of Lonergan and others like him 'wisdom has built herself a house' (Pr 9:1) in which the new holistic paradigm can flourish.

The convergences that are both possible and already taking place can be indicated by referring once more to the dialogical work cited above. Capra[12] and his Benedictine interlocutors are working with five features of the new paradigm, designed to bring both science and spirituality together.

First, there is the shift from the part to the whole. In contrast to the atomistic, analytical approaches characteristic of Bacon, Newton and Descartes, contemporary holistic sciences prioritize the interrelational whole of any given reality. This is paralleled in theology, not only by working out more fully the interconnections of the mysteries or articles of faith among themselves, but by situating such connections in the fuller context of connections which characterize ecology and cosmology. Hence, the scope of the challenge: to connect the scientific data on the genesis of the cosmos with both the psychological data of the genesis of the self, and the religious data on the genesis of God within the world through the mystery of the Incarnation.

However, such connections are not merely formal or objective as though it were merely a matter of having a more comprehensive 'look' at reality. They are rather the outcome of a method that interconnects various methods of investigation in the dynamics of consciousness. Such dynamics are brought to mind in Lonergan's imperatives: 'Be attentive!' (empirical consciousness); 'Be intelligent!' (intellectual consciousness); 'Be reasonable!' (rational consciousness); 'Be responsible!' (moral and affective consciousness).[13] I might remark that whereas other disciplines of either the sciences or the humanities have been content to express their new paradigms simply in terms of a new synthesis either within one discipline or within a vaguely considered multi-disciplinary or interdisciplinary approach, the special complexity of religious data and the hermeneutical problems it has had to face have forced theology to come up with a unifying method of considerable refinement. Because this refinement is based in the dynamics and demands of human consciousness, it may well emerge as a considerable resource for other forms of holistic learning.

Secondly, the new paradigm expresses itself in terms of the shift from structure to process. Structures are now seen as manifestations of an underlying process or self-organisation of some kind. Theology, for its part, has begun to respond to this aspect of the shift. What have been long considered as separate structures or aspects of revealed facts of one kind or another—creation, Trinity, incarnation, cross and resurrection, the Church, the sacraments, eschatology—are now treated in a far more dynamic or processive manner. Hence, the whole range of separated theological themes has tended to become focused in terms of the one mystery of the self-communication of God to creation, as in the interconnective theology of Karl Rahner; or in the emergence of the Cosmic Christ, as in the evolutionary thinking of Teilhard de Chardin; or in terms of the universe of 'emergent probability', as Lonergan describes it in his magisterial *Insight*; or in the evolutionary-ecological analogical sweep of the Christology of J. L. Segundo. In such treatments, the self-communication of God is linked to the self-transcendence of the human in the emergent process of the world's coming to be.

While, in a sense, such thinking recovers the dynamic of the great theologies of the past (the *exitus-reditus* [emergence and return] scheme of Aquinas, and Bonaventure's more Platonic sense of 'the Good as diffusive of itself' [*Bonum est diffusivum sui*]), it goes much further. As in the scientific paradigm, the process explains the structure; and 'nature' is no longer a fixed entity but a heuristic or explanatory notion within the larger dynamics of history and evolution. More to the point, in such a shift, God is the creator and sustainer, not of a world of fixed natures, but of a world of time and process. The former 'timeless truths' of theology are now embodied in time as the meaning and momentum of the temporal world. To revert to the traditional idiom mentioned in the second chapter of these reflections, faith seeks understanding in reference to the process of realizing our ultimate destiny.

Thirdly, there is a shift from objective to 'epistemic science' as Capra notes. Reality is not disclosed by having a good look at it, however refined the optical instruments might be. The knower is involved, as a meaning-maker, in the process. The observer is part of the reality. Through knowledge reality is not so much confronted as enriched, while the universe becomes luminous to itself in those who explore it.

This aspect of knowing, so dramatically rediscovered in the early explorations of quantum physics, likewise has theological parallels. I mentioned above Lonergan's work in grounding theological method in human consciousness. His distinctive approach can be summed up in the axiom: 'objectivity is the fruit of authentic subjectivity'. The truth of things is discovered only to the degree that the full capacities of the subject are involved: in sensing, imagining, questioning, pondering, responding, loving. Here, the instructive instance is the knowing of another human being. An engagement, far more vital and intimate then just looking or listening or collecting data, is presupposed.

I note, too, that Lonergan's notion of objectivity as the fruit of fully activated subjectivity has a family resemblance to the views of the scientist-philosopher Michael Polanyi.[14] Historically, both are connected to the Aristotelian principle that the mind knowing and the reality known coincide in the one act.

Fourthly, in the scientific version of the new paradigm, there is the shift from 'building' to 'network' as the basic metaphor for knowledge. It is not so much a matter of adding to known reality bit by bit, so to speak, but of apprehending the whole in its interconnectedness. To treat anything as a bit or part isolated from the world of relationships in which it exists is to overlook its essential features. If a celebrated French philosopher promoted *distinguer pour unir*,[15] to distinguish in order to unite, as the governing principle of reflection, today the emphasis is reversed: to make distinctions within the prior, fundamental wholeness and network of reality.

Here, too, theology must be aware of its own resources. There is an emblematic significance in the fact that the fundamental mystery of Christian faith is the Trinity: the absolute is fundamentally relational; the divine persons are self-constituted as 'subsisting relationships' in the processive mystery of God. If God is intrinsically relational it comes as no metaphysical surprise to find that the universe is relational in its every aspect—a network in that sense. I will develop the significance of this in a later chapter. But beyond this all-important doctrinal focus, the development of theological method itself is more and more along 'network' lines, or, as Lonergan would phrase it, method is 'a framework of collaborative creativity'.

The former 'building block' procedure of, say, scriptural data, defined doctrines, theological opinions, philosophical reflection, is being replaced by the 'collaborative creativity' of different interdependent specializations. 'Framework' is perhaps too spatial and static a metaphor for the new ideal of interchange and collaboration. Still, it makes its point. Such a method is more apt not only to explore the realities it considers as a network of relationships, but to perform such exploration in a network of consciously assumed collaboration.

Fifthly, there is a shift from truth to approximate descriptions. Science has abandoned its quest for Cartesian certainty to be content with the more tentative, ever revisable notion of probability. The human search for truth unfolds as an ever more refined series of approximations to what probably is the case. It is not a matter of compressing the world into our scientific models, but of designing models of exploration that illumine our participation in the universal process of becoming.

Perhaps this seems the feature of the new paradigm most inimical to theological exploration. After all, faith lives from the conviction that the divine is self-revealed, and that a Yes to that cannot turn into a No. But even here, matters are not so cut and dried. First, there is the character of analogical knowledge: it was a council of the Church which defined that though theological understanding can establish certain likenesses and analogies between normally accessible realities and revealed mysteries, we must always remember that the dissimilarity existing between the creature and the creator is greater than any similarity.[16]

Further, in line with what we have already mentioned, the divinely given 'data' is always explored in terms of the limitations and opportunities afforded by a particular stage of history. Even if the Yes of faith to the divine mystery is unconditioned and without reservation, the understanding of that faith in its real bearing in the real world of our experience is always limited by our pilgrim state. There is the increase that comes from the accumulation of experience and insight. Beyond that, there is a deeper, more interior possibility inasmuch as believers become more religiously converted to the divine mystery in self-surrender, or more intellectually converted as they drop the deficient constructs of imagination or systematic meaning for something more critically established; or more morally converted as new values draw them into greater responsibility and compassion. Then, too, there is the possibility of a more fundamental enrichment of knowledge as consciousness becomes more differentiated in its capacities, skills, tastes, feelings, categories and language.

All this is to say that the radically inexpressible truth of faith is quite compatible with the successive approximations that history allows. In other words, doctrine develops, and the 'hierarchy of truths' is reshaped and re-expressed to find expression within the meanings and values that inform a given culture. Hence, while faith can and must treasure its intimate experience of God-given truth, the defensive rationalism that looks for certitude in each and every instance is deeply inimical to real learning. Theology can be quite content, then, to join with other forms of thought that are less concerned to trumpet forth the certitude of their attainments, and more inclined to accept the humbler role of exploring the meanings with which the universe is illumined.

Though these remarks on the new holistic paradigm and theological understanding are necessarily general, I offer them in the hope that they will underscore the possibility of reciprocal enrichment—not of mutual antagonism—in the one exploration of the meaning of our existence.

4. The New Age movement

Our reflections on the new inclusive story, on the new paradigm of learning and on the relevance of both to Christian theology would be somewhat innocent if no reference were made to the vast, confusing, often contradictory phenomenon of the New Age movement as it has developed over the last twenty years or so. Though the bewildering variety of its theories and practices suggests no overall synthesis, a number of worthwhile emphases stand out in its representative literature.[17]

I would suggest the following nine. Though in many ways I am in agreement with them, I will add a critical note to each to remind myself at least of the difficulty of getting beyond a general emphasis to a critical position, from which constructive dialogue can become possible.

i. In the first place, we find a consistent effort to counteract a great range of inherited dualisms in the Western intellectual and spiritual tradition, especially as these were further intensified in the Enlightenment. Hence **a new cosmology of wholeness** rejects the separations of science and religion, of body and spirit, of matter and consciousness, of thinking and feeling, of male and female, of transcendence and immanence, of objectivity and subjectivity.

The general problem here, despite the welcome stress on holism, is that *dualism* is often little distinguished from *duality*. The former implies a distinction in which one element is made ideologically subordinate to the other; whereas the latter is simply an expression of the capacity of human intelligence to distinguish various aspects, or principles or polarities in existence. Often such distinctions have been remarkable attainments in the history of thought. The classic example is the duality of body and soul in the constitution of the human being. Hylomorphism understands the material and spiritual to be related in such a way as to constitute the integrity of the embodied, animated, unified reality of the concrete human being. Such an Aristotelian and Thomistic position represents a great advance over the *dualistic* Platonic tradition which demeaned matter and body in its exaltation of the spiritual. Hence the danger in this emphasis of 'New Age' thinking, of sacrificing the constitutional richness of reality to an undifferentiated confusion: 'No one knows better the true meaning of distinction than they who have entered into unity' (Tauler).[18]

ii. In contrast to the doctrine of the transcendence of God over creation, and of the infinite ontological distance between the creator and the creature, New Age thinking celebrates the **immanence of the divine within all things** as the ground for a universal interconnectedness.

Again, as we will stress later, this can be a healthy recovery of a genuine sense of the divine presence within all reality. However, the danger is that God is made either identical with what he has created, or is conceived of as part of creation itself. In contrast, the great tradition of theological and philosophical thought considered that the reason for God's unimaginable intimacy with creation lies precisely in his transcendence as the source of all being. Divine transcendence is not opposed to immanence, but is the reason for it. Confusion here draws the accusation that the 'New Age' is a regression to paganism and pantheism, when what may be struggling for expression is a venerable form of *panentheism*: all things existing in God, as their source, ground and goal. Though in, say, the Thomistic version of this tradition, God is not the undifferentiated potentiality of prime matter, nor the soul of the created world, nor the universal process itself, the processive life of the Trinity is the dynamic matrix in which the universe comes to be. We shall devote a later reflection to this point.

iii. Then, too, an **intense optimism about individual and social transformation** characterizes this movement. The new age is possible; human potential is limitless; its dynamics are beckoning and available. Emphasis on sin and evil, on redemption and conversion must yield to a world of 'original blessings' in a good creation.

Now it is possible to delight in the fundamental goodness of creation and the priority and universality of grace yet still demand a great realism concerning the tragedy of the human condition. The closing decades of this especially tragic century hardly permit any innocent optimism. Techniques of meditation and heightening of consciousness, exercises in transpersonal awareness, can be elaborate subterfuges of denial if the deadly reality of the 'heart of stone' is not faced.

I suppose it comes down to distinguishing between optimism and the transcendent character of hope. A hope-filled apprehension of the universe is a high religious achievement. It is attained in the teeth of the 'problem of evil', in a gracious acceptance of the seemingly random, entropic and humanly incomprehensible complexity of the universal process. The rhetoric of Christian tradition which would speak of the creation and the fall, of nature and of grace, of conversion from evil and conversion to the good, of the reality of human freedom blossoming only through surrender to the divine, of the cross and resurrection, of the Pauline conflict of 'flesh and spirit', is evidence of a far more nuanced sense of the human condition than a technique-induced New Age consciousness.[19] Still, to say this is to state the conflicts rather than the possibilities of deeper learning on both sides of the divide.

iv. Less ambiguously, New Age thinking stresses the **planetary dimensions of human consciousness** and of ecology itself. The particularities of culture and history, often so defensive and divisive, have to yield to the global dimensions of coexistence. National or even international movements still carry the baggage of particular interests and restricted notions of the common good.

The main problem here is the way in which this planetization occurs. If it takes place as a genuine ecological communication between all the differentiations of consciousness and all the varied, long-term achievements of human culture, such a convergence of experience, thought and feeling would indeed promise a new global conscientization. If, on the other hand, such a planetization becomes something rather less than this, and more like a marketing of spiritual self-help techniques, a new spiritual consumerism, a new imposition of North American or European post-modern pathologies, New Age planetization will be an oppressive element in the ecology of a global culture. Transcultural communication, cosmic interconnectedness has to be something more than standardized enlightenment.

v. Then, there is a sense of the **cosmic pervasiveness of the evolutionary consciousness**. An analogical imagination links everything in a pan-evolutionary movement. Everything and everyone connects in a universal becoming.

Clearly, these reflections are written in obvious agreement with such a sense of reality, as is so much of the leading science, philosophy and theology of our day. The dark side to this evolutionary awareness is the questions it leaves unanswered, above all those of an eschatological nature. Is the present and the past devoid of any meaning save in reference to our present apprehensions of the future? Was all the suffering of the past just raw material for the New Age? Is there no value in the present save in its evolutionary potential? What of the frustration, failure, waste, tragedy, extinction of species, death of individuals, and the incomprehensible randomness of both the past and the future? Does being vulnerable to all that realm of entropy leave human beings with any meaning at all? Finally, does a simple New Age evolutionary optic tend to place its possessors outside the actual evolutionary process, at the expense of immersing them in it; thus to excuse them from the demands of patience, surrender, historical memory, and of hope against hope? An evolutionary ideology, contrary to evolutionary science, has a strange way of offering both instant success and a kind of infused knowledge on whither things are tending.

vi. More negatively, there are those **influences in the history of the West that New Age thinking rejects**. First, it unwearyingly points to the deficiencies of establishment Newtonian science and Cartesian anthropology. Such an inheritance narrowed the whole scope of human experience and diminished the range of alternative modes of perception. The human subject, in the full panoply of consciousness, has been notoriously absent. Then, there is a general disaffection with 'institutional' religion and with Christianity in particular. The more institutionalized such religion became, the more doctrinal its formulations, the more it walled up the sacred in its own mediations and rendered religious experience inaccessible to its people. With its doctrine of sin and the need for redemption, it demeaned the body and creation, and minimized human potential. Its post-mortem promises further alienated its adherents from the material cosmos. A transcendent God from beyond the world offered a precarious salvation to spiritual souls imprisoned within it.

Unhappily, such a description is not entirely a caricature. And I, in common with the generality of theologians and philosophers today, cannot but endorse a 'return to the subject', and a reappropriation of the whole scope of human experience. On the other hand, matters are more intricate than a leap from one mode of knowing to another without questioning why such agility is justified or possible. Newton's fervent religiosity is well known; but the case of Descartes is perhaps more astonishing to the modern holistic mind. It gives us pause to recall that this super-rationalist, super-individualist, mathematical French *philosophe* considered that his discoveries were the result of a visionary experience inspired by Our Lady. In such a conviction, he vowed to make a pilgrimage of gratitude to Loreto—which he performed in 1622.[20] Apparently, Descartes was more complex than Cartesianism.

So, we are led to ask, what was really going on then? And, what is going on now? Apart from referring to remarks made above on historical developments, differentiations of consciousness and levels of conversion, I might just add one simple point.

It comes down to this: when we attempt to evaluate the past, the actual richness of past experience is largely inaccessible to us now. What we are dealing with, unless we are gifted with extraordinary historical imagination, are very narrow objectifications of what was going on in those days and in those now dead generations, be they Counter-Reformation Catholics, Newtonian scientists or Cartesian philosophers. Doubtless, the current holistic mentality has developed a much greater refinement in attending to the data of experience and naming it. Still, the real challenge is not one of simply dismissing the narrowness of past achievements, but of integrating all the accessible data into a comprehensive method which does justice to the complexity and richness of reality.

It is not enough to have a new guiding symbol of wholeness, nor a more inclusive feeling for the whole, nor even a new scale of values and range of meanings. Some thoroughgoing integration is required, by which we can detect the lacunae or distortions in the past—and in ourselves. Otherwise the promise, say, of holistic medicine will be stopped short in its inability to find an answer to the HIV virus; and the desirability of a mystical religious experience will be unable to relate to the necessity of an institutional expression of religion strong enough to confront the institutionalized social, political, economic distortions of culture.

In short, a holistic mentality can only be sustained by a holistic method. Otherwise it will find itself repressing or overlooking whole ranges of data just as effectively as anything in the past.

vii. Moreover, New Age thinking enthusiastically embraces '**the new physics**', the universe of quantum and holographic perception, as more hospitable to the phenomenon of consciousness and transformation.

I agree; but I still think it is worth noting what is going on in terms of belief. Our culture is deeply affected by the scientific myth. The only ultimate authority, the only sure way to knowledge, is science, or what passes for it. Our world picture is formed, often enough uncritically, by what are taken to be the findings of science. It is difficult to keep in mind that genuine science is an ongoing, approximate procedure, employing, in its technical expression, mathematical symbols of extreme sophistication, compared to which any prose account is like the wildest kind of poetry. Often enough we are left in the position of being simple believers in what they, the scientists, *really* know. A consequent mistrust of our own firsthand experience of the world, especially in its religious or artistic or moral dimensions, is a growing possibility. The result is that it is more socially acceptable to believe in the Big Bang than in God's creative Word; easier to believe in the billions of years of cosmic emergence (recent literature speaks of twenty or most commonly fifteen, or eleven billion years—what's a billion years one way or the other to us simple believers?) than in the eternal Now that holds all time together; easier to believe in cosmic events taking place in the first trillionth of a second than in a Providence reaching into every detail of our lives; easier to believe in the hundred billion galaxies extending beyond our Milky Way than in the value of answering the needs of our neighbor; easier to believe in the eighteen variations of quarks than in the value of justice or forgiveness.

We simple believers are more culturally open to refinements of relativity theory, however much it seems to disorient our perceptions of the real world, than to the testimonies of prophets or mystics. One can more easily go along with the fundamental four forces unifying the physical universe than in the disclosive power of love; just as one can tend to be more impressed with the authority of the New Physics than any religious or moral tradition, however venerable and ancient. Perhaps I am more spontaneously suggestible to the latest findings on Black Holes or chaos theory or dark matter than to the reality of Christ's resurrection. Popular scientific culture prepares us more for amazement at the genetic connectedness of all life than for pondering the cosmic significance of trinitarian relationships. Cultural belief runs more readily with the incredibly intricate abstractions of the mathematical symbolization of the universe than with the Word who became flesh and dwelt amongst us.[21]

Hence a critical, extremely complex issue emerges. Belief in the authoritative findings of others is certainly a way of participating in the great historical and social process of coming to know and explore the manifold meanings of the real. Yet there are many kinds of belief: in the scientific case, we rely on the explorations of research; in scholarship, we take as worthwhile the judgment of reputable thinkers; in art, we accept the productions of the great spirits of the age; in religion, we accept the witness of the prophets, mystics and doctors of the tradition; in everyday life, we are surrounded with proliferating expertise.

Still, none of these instances of belief necessarily excludes all or any of the others. Nor do any or all of them excuse us from a personal responsibility. Beliefs can be mistaken, or need revision—a feature of believing that comes to light in the course of time. A usual sign of the authenticity of belief is its ability to speak to, or allow for, the whole human condition as a manifold reality. It is at that point, I would suggest, that each of us, whatever the promise of a holistic mentality or a New Age transformation, must maintain a critical reserve. Any act of belief must send us back to our responsible selves.

viii. A further valuable feature of the New Age mentality is its more or less **mystical sense of the universe as a uni–verse**, a whole, and all, long preceding the specialized analyses and divisions of sciences and all the different ways of knowing.

Such a holistic anticipation of the whole, such an inclusive presentiment of the all, has a long philosophical pedigree. The dynamics of human intelligence are powered by the transcendental notions of the one, the intelligible, the true, the good, the beautiful. Categorically, the mind attains to the particularity of this or that object only within an horizon of the totality of being.

Serious contemporary questions arise concerning the nature of this unity. The manner in which Teilhard de Chardin formulates the issue documents not only something of his own struggle to formulate his own sense of universal event.[22] It also leads us to a distinction of vital importance, if we are to arrive at a genuinely ecological and communal sense of existence. Here is Teilhard's description of two kinds of mystical road:

> The first road: ... to come together with a sort of common stuff which *underlies* the variety of concrete beings. Access to Aldous Huxley's 'common ground'. This procedure leads ultimately to an *identification* of each and all with the common ground—to an ineffable of de-differentiation and de-personalisation. Both by definition and by structure, this is a mysticism WITHOUT LOVE.
>
> The second road: to become one with all ... no longer by 'dissolution' but through a peak of intensity arrived at by what is most incommunicable in each element. This procedure leads ultimately to an ultra-personalizing, ultra-determining, and ultra-differentiating UNIFICATION of the elements within a common focus, the specific effect of love.[23]

In contrast to the primal, pre-biological and pre-personal unity of the first road underlying all things, and into which everything ultimately dissolves, there is the unity which is that of union with the ultimate, and communion within it. Differentiated realities are not dissolved but brought to their interconnected completeness in being. The ultimate unifying force or energy is love, the affirmation of each and all in its irrepeatable particularity. From this love there flows an intimate knowledge of the other; indeed, in Teilhard's understanding —and this accords with the general tradition of Christian theology— such loving knowledge is a participation in the creative knowledge and love of God, the source and goal of the universe of particular realities.[24]

The attraction of 'Eastern', monistic mystical philosophies can be easily documented in much New Age and New Paradigm thinking. One is left to wonder whether those who follow such a 'road' realize the full import of the choice they have made. At first glance, it would seem to undercut the realism of any ecological commitment, to say nothing of the quality of the transformation of the self that it promises. At least, the necessity for dialogue is evident: a deeper appropriation of the 'Western' tradition (as, say, it appears in Aristotle, Bonaventure, Teilhard and Lonergan) may well reveal an abundance of overlooked resources.

A mystical resacralization, or re-enchantment of 'nature' or science, called for in reaction to the nihilism and mechanicism of recent centuries, will mean something vastly different for those who travel the different 'roads': for the first it is a path to dissolution into a formless 'all', a unity without difference; for the second it is participation in a world of real relationships, an ultimate unity-in-difference.[25]

ix. Finally, **New Age thinking relativizes the centrality of the human.** Ecologically speaking, the human is one of millions of living species. Cosmologically, the human is a phenomenon occurring within an emergence of cosmic proportions. 'Anthropocentricity', biblical or otherwise, is on both counts a distortion.

Since that burning issue will be the subject of an extensive later reflection, I simply point once more to the complexity of the issue: judgments on such a real or possible distortion remain human judgments. They are made within human consciousness and are communicated in human speech. Desirably, the relativization of the human indicates an option for humility within the vastness of the cosmos, and for a more profound relationship of care for other species of life that humans have so unthinkingly exploited. The 'galloping case of Enlightenment epistemology' needs to be redeemed into a more tender, participative, indwelling kind of knowing.[26]. Possibly, however, extreme criticisms of anthropocentricity mask an effort to escape from the human, to evade the issue of human culture within the ecological commonwealth, and to demean or overlook the role of the human in the unfolding of the cosmos itself. If that is the case, then, unwittingly or deliberately, anti-anthropocentricity is calling for a kind of decapitation of evolutionary process and a suppression of the consciousness in which the wonder of existence emerges.

I hope, then, that these nine emphases, typical of New Age and

New Paradigm thinking, are a useful enumeration of points for dialogue. Admittedly, for any of us who venture into such exciting new perceptions of reality, our explorations can easily be dismissed as eclectic or amorphous. And yet, there is evidence of a deeper search for meaning, a desire to 'learn by heart' the movement and form of the universe in which we so uncannily exist. True, the result is often a strange conglomerate of astrophysics and astrology, of the occult and the deepest mystical and philosophical perceptions. Still, when a flatly secularized culture has so repressed our deeper perceptions and experience, these will out, in forms sometimes odd and unnerving.

Perhaps the soundest tactic of all is to keep trying to get to the question that is being asked in each instance. When traditional doctrinal and conceptual forms are no longer available as answers, it becomes a matter of learning all over again, by heart, what is present but too long unnamed. It can indeed by a benign conspiracy when deeply felt human values are asserting themselves in various contexts of science, economics, politics and ecology, to say nothing of religion itself. Such is the case, too, when, in these contexts, experts and amateurs can meet. The experts can find themselves disarmed in the realization that we are all amateurs, both in regard to our shared humanity and in relationship to its place in the ever-mysterious universe. On the other hand, an amateur status in science can find new heart as the wonder of the universe is disclosed in the sophisticated techniques of modern exploration.

5. Pathological holism

The situation admits, of course, no simple-minded, one dimensional response. If a new holistic vision comes to mean some form of exhaustive explanation of all that is and lives, a sense of oneness is coming dangerously close to being merely 'one point of view'. The reaction to such a reduction is predictable. For example, it is found in the neuralgic repulsion many feel to what they apprehend as doctrinaire 'Greenness'. In a century that already knows enough about totalitarian pretensions, any wisdom that has resulted or that remains is extremely sensitive to even a whiff of a new kind of fascism.

Dread of the future, panic about the present, the traumas of confrontation coupled with a pervasive sense of crisis—all could easily turn into one more unredeemed struggle for power, one more fanaticism, one more fundamentalism, one more ideology rigidly excluding the whole mystery of our shared life.

The backlash would be predictable. Some critics already protest against a supposed Leftist takeover of the Green movement. The other extreme sees it as a capitalist manipulation of poorer countries to keep raw materials safe and investment potential secure.

Indeed, some ecological statements seem to come from the depths of a weird self-hatred, as though human beings were some kind of foreign body in the life of the planet, a cancer on its growth, a mutant ugliness in the perfection of unspoilt nature. From such rather raw feelings arises much of the reaction against 'anthropocentricity', especially when such a reaction is innocent of the realization that it is only within human consciousness that our earth experiences either wonder or horror, or flowers to creativity and care.

Then, too, in the name of ecology, the ecology of our human communion on this planet is compromised. Immigration programs are restricted, the status of refugee is more and more narrowly defined and the urgent problems of overpopulation occasion brutal, racist solutions. Humanity seems to have the special ability of turning against itself with a savagery that would never be countenanced in dealing with other species. Self-hatred has never been productive of anything; and whatever the problems we face, offering human sacrifices to an idolized 'nature' is not the solution we want. We are not ghosts haunting a world machine, nor parasites in a vast organism; nor, for that matter, rangers in a great national park.

The use of the word 'ecology' does not necessarily put one on the side of the angels. Hence, beyond political and economic suspicions regarding the real intent of ecological interventions, somewhat deeper questions will stir: is the ecological movement a regression to primitivism? Is it basically a hatred of the human? Perhaps a resurgence of pantheism? Is it the final spasm of the West's cultural death-wish? True, we may dismiss such interrogations as unworthy or misplaced, but only after they have been allowed to resonate in our deepest conscience.

How then are we to relate to nature 'in the round', in a 'down to earth' ecological manner? We cannot escape from our human responsibility, just as our humanity cannot be removed from earth. To attempt either such escape or removal would be inherently destructive.

The dangers are sufficiently real to make it imperative that various visions of wholeness interact in a complementary rather than in a mutually exclusive manner. We human beings are now being drawn into a time of collaboration as the necessary condition for any global history. The ecological moment is not a time for crass adversarial warfare. One must be a conservationist not only of the variety of nature, but of the values of culture that nurture the human spirit to a sense of wonder, gratitude, hope and humility. If the nature of ecological issues is to inspire a huge collaborative multidisciplinary effort extending over many generations, patience becomes one of its essential virtues.

We might note in passing that Christian faith has a distinctive historical experience to call on in this conflict between an ecological vision and reaction to it, between the new holistic construction of meaning, and the deconstruction that disallows any definitive solution.[27] Christian history has faced such problems. On the one hand, there is the one truth of one creator, the universe made one in the cosmic Christ in whom, through whom and for whom all things exist: this is the fundamental 'Catholic' vision of wholeness—a universe, a seamless garment, apprehended in one 'analogical imagination'. It extends the biblical 'Wisdom tradition' which attempts to bring faith and culture, grace and nature, history and cosmos together.

On the other hand, the end is not yet; 'we do not yet see all things in subjection to him' (Heb 2:8); nature still groans in its eager longing for liberation (Rm 8:19); 'It has not yet appeared what we shall be' (1 Jn 3:16). The pilgrim path is still a long one, with all kinds of dread self-dispossession, self-surrender, of dying into something more, something always greater.

Here we hear the voice of the reformer, the 'Protestant', the prophet, contesting any foreclosure of vision into settled system. It demands that wisdom remain a search, an overture to what immeasurably transcends any present attainment.

That is the 'common sense' which Christian experience, in its present ecumenical phase, has taught us. It may not be irrelevant to the conflict between holistic thinking and deconstructive suspicions in the present 'post-modern' era.

Whatever the case, in the dialogue now possible between the various *logoi* of theology, ecology and cosmology, we are learning to word the teeming wonder of the world, on the way to becoming more responsible participants in the mystery of life's unfolding, and more inspired celebrants of the marvelously diversified cosmos in which we are embodied.

Education to a more wholesome ecological sense of reality has become a cultural imperative. Pope John Paul II writes:

> An education in ecological responsibility is urgent: responsibility for oneself, for others, for the earth. This education cannot be rooted in mere sentiment or empty wishes. Its purposes cannot be ideological or political. It cannot be based on a rejection of the modern world or on a vague desire to return to some 'paradise lost'. Instead, a true education in responsibility entails a genuine conversion in ways of thought and behaviour.[28]

If an integrated ecological education does not take place, one fears that not only will the Christian sense of the whole be stunted, but the ecological movement may slowly congeal into an oppressive kind of gnosticism.

I cannot believe that we are very serious about the crisis that is upon us. One must suspect the presence of a deep cultural unwillingness to face the pain and the cost involved in the about-turn we need to accomplish. For example, one of the keenest agonies human beings can suffer is that of the 'withdrawal symptoms' associated with trying to break some dependency or addiction. We all know well enough that there can be no solution in the simple optimism that expects it will all work out with a minimum of struggle. Such an attitude is a denial of the extent of our problems, and repressive of huge anxieties that threaten. The only way out is a larger shared imagination,[29] in which the confusion and the panic can be transmuted into both a collaborative effort and that keener dependence on the sustaining power of ultimate Life on which all hope is based.

For ecological consciousness is now awake to the necessity of a *metabolic* relationship of human kind to the earth. The sorry fact is that our present relationship to biosphere is more aptly described as *diabolic*—disruptive, sundering connections rather than respecting them.

The emerging conflict has sparked the necessity of a new *symbolic* (bringing together) sense of human existence in the living world and within the universe as a whole. Religious faith, most of all, is expressed in the great governing symbols of existence. They disclose ranges of reality broader, deeper, more ultimate and more inclusive than a purely functional apprehension of our world. The symbolic 're-members' the scattered, distracted fragments of our being in the world, by intimating the mystery of it all, the healing wholeness of reality revealed as gracious. Only in a symbolically disclosed horizon can the diabolic dismemberment of our relationships to the world and to one another find its exorcism and eventual healing in a genuine metabolic relationship with the biosphere. For this reason, reflections on the great symbols of incarnation and Trinity, of eucharist and sexuality will be later investigated as the foundation of the hope we most need.

The emergence of a new paradigm, not only for science, but also for life in its concrete unfolding, has become a necessity. Our culture has drifted far from its original wisdom, and is in need of a thorough–going new development. Some speak of 'a pervasive cultural pathology'.[30] The sobering point here is this: if we don't realize that the malaise is first of all one of the inner ecology of our thought, of our feelings and values—that is, of culturally mediated reality—we will be numbed in our abilities to develop the larger wisdom of a fuller, more inclusive life together, in the one planetary biosphere. There can be no remedy for the threatened environment save by paying attention to the quality of human consciousness, and the meanings and values it expresses or pursues.

A mutilated or stunted inner environment is not the best resource to bring to the problem. To act without realizing how polluted our inner environment has become is to infect the wound further, rather than to heal it. Without a radical cultural transformation,[31] the threatened ecology of the planet will be addressed only with the resources of a stunted inner environment: ecology will be merely another theatre for an unredeemed power-struggle. An habitual attitude of rape in regard to the earth is not easily changed into the tender exchanges of love. Again to quote Pope John Paul II:

> Modern society will find no solution to the ecological problem unless it takes a serious look at its lifestyle. In many parts of the world, society is given to instant gratification and consumerism while remaining indifferent to the damage these cause... Simplicity, moderation and discipline, as well as a spirit of sacrifice, must become part of everyday life, lest all suffer the negative consequences of careless habits of the few.[32]

True, there are some who would see human consciousness as a miasmic mist floating on the surface of the planet, and so the cause of all the poisoning and destruction of nature. At the other extreme, others so glorify this most mysterious self-presence of the human that they forget it is embodied, 'earthed' in the given physical and biological processes of the planet. Between saying that we have been anthropocentric to the neglect of a larger biocentrism, and the seemingly obvious fact that it is only through human consciousness that the ecological state of the planet is perceived as a problem, lies a world of great complexity.

Indeed, in a world of threatened species, the distinctiveness of the human also faces the perils of extinction or diminishment. There is an alienation, not only from without (depleted resources, poisoned air and water and earth and forests) but from within, given the violence and hatreds that lie close to the human heart. There is no need to document what we have done to one another on the individual, familial, social, national and regional level.

Has humanity really given itself a chance—at least on the scale on which we are called to live? What can liberate us, redeem us, into the attainment of our best selves? That might be the new form of the question in the current situation, when the human species is in danger of lapsing into a form of self-hatred. Alienation from our biosphere and ourselves can only be remedied by the more critical self-appropriation of the best in ourselves in terms of art, intelligence, morality and faith.

One aspect of the dignity and drama of human existence is that it is that unique mode of being in which both the mystery and crisis of the world come to awareness. One feature of that self-awareness is its exuberance: the human is related to the millions of other species, embodied in a cosmic matrix of elements and dynamisms that make life possible. Yet that exuberance is tempered, if not appalled, by a sense of crisis. The damage human history has inflicted is recorded in the extinction of thousands of species with which we once shared life on this planet.

This unique human awareness, as it blossoms into a sense of inclusive solidarity with the earth, is becoming newly alert to the myopia and addictions that have made the human presence on the planet a major ecological threat.

We are challenged by the demands of a profound and laborious conversion. Overcoming any addiction is a terrifying prospect. To break away from the accumulated and deeply embedded cultural patterns of an anti-life consumerism is going to produce the trauma of agonizing withdrawal symptoms. Though in a world vastly different from that of one of Israel's prophets two and a half millennia ago, the call to a change of heart is even more piercing and urgent:

> I will take you from the nations and gather you from all the countries, and bring you into your own land. I will sprinkle clean water upon you, and you shall be clean from your uncleannesses, and from all your idols I will cleanse you. A new heart I will give you, and a new spirit I will put within you; and I will remove from your body the heart of stone and give you a heart of flesh. (Ezekiel 36: 24–26)

That prophetic expression of God's promise to the exiled, scattered, dispirited people of Israel can rightly be heard by all the present peoples of the globe in this moment of crisis. The deadly isolation of the heart of stone has to yield to a more cosmically attuned heart of flesh. Our idols must fall and our demons be exorcised if ecology is ever going to mean something.

Thus, the ecological turn in consciousness has come to mediate a deep criticism of modern culture and the determination to find something better. The *logos* of ecology has come to stand for a larger logic, a larger wisdom, in a new kind of economy, a new kind of politics and world order. The kind of questions facing human history have been well phrased in the following passage:

> Within the economy of God, economy, ecology and *oikumene* are concerned with peculiar questions. The question of economy is: will everyone in the household get what it takes to live? The question of ecology is: will nature be given its rights, or must it protest by dying, thereby cutting off the existence of human beings? The question of the *oikumene* is: will the world become

> mutually habitable by the peoples of the earth? Will all people be able to live in the world as a home? Thus, the traditional content of economy has been *livelihood*, of ecology, *symbiosis with nature*, and of *oikumene, mutually recognized and supportive habitat in peace*.[33]

We might also add further 'eco-cosmological' questions: how does life and the 'home' belong to the universal cosmic process? Is the universe friendly to life and human consciousness, or fundamentally indifferent to it? Without a range of reference to the cosmic matrix of life and its conscious manifestation, without a sense of the original mystery out of which the 'whole' and the 'all' emerge and interconnect, ecology becomes merely eco-centricity, rather similar to the anthropocentricity to which it is so averse.

6. Models of ecological action

Responses to the ecological crises of our day are many and varied.[34] Here I intend to consider three types of response: the pragmatic, the ethical and the aesthetic. Though dealing with types always tends toward caricature, it is instructive, it seems to me, to note how each attempt to objectify what ecology deals with—the community of living things—ends by demanding a greater human participation in it.

a) The pragmatic approach

This first approach is the most accessible. It tends, predictably, to be impatient with any talk of such arcane matters as holistic consciousness, of a religious sense of creation and incarnation, or of the larger disease of culture itself. The well-intentioned pragmatist might say, 'By all means, say your prayers, and love your neighbor; but real action consists in concentrating on recycling and composting, on conserving fossil fuels, on protecting endangered species, in lobbying business and government, in getting a good educational program going, and in supporting various activist groups.' Such vigorous pragmatism objectivizes the issues as environmental. It aims to replace a bad environment with a good one, in which energy is conserved, the variety of life preserved; one, in general, in which things work better.

Yet, despite the energy and growing expertise of this pragmatic approach, it shelves a number of radical questions. For the pragmatic confront issues as problems to be solved, in the domain of 'out there'. The very vigor of the intervention makes it unlikely that underlying issues will be noticed, let alone diagnosed as a deeply distorted way of life. Activists are cast in the role of problem-solvers, as experts acting from outside the living systems they aim to protect. If there is any truth in the wry observation that those most dedicated to the solution of the world's ills are often struggling in matters closer to home, it is especially evident here. The bias of the pragmatist and the problem-solver often hurries them past the deeper challenges of intimacy with the other. For this other is typically envisaged as a problem awaiting an energetic, expert solution. To connect with reality in some other way, let alone as with an englobing mystery inspiring a deeper relationship of receptivity and participation, seems to blunt the edge of resolve.

Regrettably, the purely activist role tends to hurry past our best resources: contemplation, reflection, compassion and hope. To speak with a higher pragmatism, we can say that there is no substitute for putting our best selves into our work. But that 'best self' is the elusive resource. For example, once someone has had a stroke or a nervous breakdown, it is usually not enough for the fragile sufferer to be greeted with the resounding summons, 'Pull yourself together!' Now that presupposes resources that are simply not there. Suddenly to reverse the state of breakdown, or to become conscious of an underlying diseased way of life, are not practical possibilities. By extension, to think of the ecological crisis merely as a technological problem to be solved by better techniques leaves the whole character of technological culture unexamined; and thus doomed to repeat the old mistakes, even if less lethally. If we still think of reality like a big machine, then at best we are engineers; at worst little cogs within it.

Now, of course, the army of those who are working for better technologies, for a more healthy environment, for the conservation of endangered species and so forth, certainly deserve every commendation and support. But that is only part of the picture. A deeper involvement of the self is called for; and that comes to expression in the domain of ethics, in the consciousness of moral values.

b) The ethical approach

On this level, moral values are perceived as necessarily motivating ecological action. New inspiring, more inclusive forms of the 'common good', of 'natural law', of universal compassion, surface and find expression in the common conscience. One dominant version of such an ecologically attuned conscience is the language of rights—the right to a healthy environment; even the rights of a given environment, or of living elements within it, of animals, of rain forests, of ocean reefs.[35] However historically odd the language of animal rights and liberation must appear to be, there is no doubting the genuine, freshly perceived values it expresses.

Still, a certain reserve on the long-term usefulness of 'rights language' is justified. The problem again turns on the peculiar externality of the ecological relationship implied. For example, campaigners for the rights of those others—ranging all the way from native cultures, to animal species such as whales, koalas or possums, to biosystems such as ocean reefs and rain forests—can be so uncritically secure in their own liberation. To put the matter baldly: the 'others' are being offered the freedom already in possession, so to speak, of the liberators. At this point, it is scarcely noticed that the ideal of justice presumed in all this is itself rather narrow and distorted. It so often communicates a notion of rights that is militantly adversarial in tone, individualistic in content, self-assertive and competitive in style. It is born of a mistrust of consensus, of genuine conversation and of the complexity of the issues. The malaise of any number of modern democracies provides ample illustration here.

The language of modern justice, increasingly detached from any gracious worldview, is the rough language formed out of a violently competitive society. It is not the language of communion, conviviality, mutual belonging and humble service. Here, too, the 'medium is the message'. A morally ecological language certainly needs to be supported by a renewed sense of justice; but, as I see it, it will be closer to the imperatives of an other-directed love.

Indeed, one could argue that a justice detached from an inclusive sense of larger common good has been the cultural source of our present troubles as it privileges the individual and favors those most in a position to exploit the system to their advantage. In an imperfect world, one must concede, such a notion still has its point; but not to look for something else, a language envisaging a world of communion, of self-dispossession for the sake of being with, and for, the other, is to make the present ethical concern a harsh intolerant moralism, or another version of justice for the few at the expense of the many.

In practice, it needs the illumination of the artist and the mystic, of a larger philosophy, of a theology of reconciling hope. If conflict is not for the sake of a deeper conviviality, it will be aimlessly disruptive. However vulnerable and utopian it will always appear, the supreme value of love is the radically real issue: the love of one's neighbor and the neighborhood; of loving one's enemy and what is strange, untamed, hitherto unnoticed or 'useless'; loving it because it exists, given into a world of being and life beyond any calculation of one's own, already there, with its own past and future; and offered to us in a holy communion that precedes any distancing or antagonism. Such love is an attitude of welcoming and hospitality within the mystery of creation.

On the other hand, a narrow justice as the requirement of everyone having individual wants filled and being treated as each feels he or she deserves is rather close to a Christian description, not of ecological communion, but of hell.

c) The aesthetic approach

Nor is it a matter of aesthetics, at least in the superficial sense of human beings becoming planetary landscape gardeners. The destruction of what was once beautiful or the beauty of what can easily be destroyed certainly provokes its own powerful reaction. But nature is more than an inspiring ambience: concentrating on beauty alone, even the beauty of all the varied creation of plant and animal life, is not enough. For one thing, we are positioning ourselves and a good deal of nature itself, out of the picture. We are seeing ourselves as vacationers, tourists, spectators, park rangers, landscape gardeners who want to preserve what we imagine has to be a beautiful environment. This can be the most subtle form of ecological insensitivity, in which a weed is anything I have not planted, and a pest is any creature that interferes with my plans. We are denying the full given reality of nature. And that includes death and decay; and more threateningly, plague, disease, natural disaster.

As every beautiful woman has always known, to be treated merely as a beautiful object can be a belittling experience. Her essential humanity, including all the negativity and struggle implied in that, is being repressed, disallowed, denied. To be treated merely as an aesthetic or sexual object is an insult to her distinctive personality with its aspirations to relationship and growth. And, of course, beauty is notoriously in the eye of the beholder. For some the Finger Wharf in Woolloomooloo is a seventy-year-old eyesore obscuring the natural shore of the Sydney harbor; for others, it is beautiful, a part of the place, something to be cherished, and in the event, defended. So to leave ecological questions on the level of a pleasing environment is a hugely repressive denial, ultimately confusing and destructive.

A deeper and more realistic kind of ecological approach must unfold in a larger horizon. Its posture will be one of openness to the whole mystery of nature in all its grandeur, beauty, struggle, tension. Any activity will come out of an intimate receptivity to the whole as something to belong to, live with, work for, yield to; yet in a way that allows for the human contribution, the distinctive mutation that human beings can bring.

How, then, can we think of this transformative activity on the part of the human with regard to the world of nature? It has been recently suggested that the human is related to nature less in terms of the contemplation of beauty than in the demands of the artistic creation of the beautiful.[36] Artists have a respectful working-relationship with their materials. Within given limits of the stone or paint or tonal scale or physical movement, a creative interaction occurs. A new form is educed. Reality is illumined by the human creative act. Matter is newly imagined, released to a new intensity of existence, a new form of embodiment.

Such an approach allows for the creative activity of the human, but only within given limits that have to be respected. To this degree, there is a certain co-realization as nature plays itself through artistic imagination into a life-nourishing beauty.[37] The role of the human implied in such a relationship is that of being the medium, as it were, through which nature can be elevated to a new form. In human imagination, nature stands ready to receive the inexhaustible, the surprising, the revelation. Through art, nature attends on the transcendent.[38]

Now, this approach bears some analogy to the venerable medieval conviction of 'grace healing, perfecting and elevating nature', in which 'grace' is understood as the loving excess of the creativity of God, and 'nature' is limited to human potential. In an ecological horizon, we might understand the human as called to be grace in regard to the larger domain of nature; thus, to heal, preserve, perfect and elevate it. In the narrowly theological tradition we refer to, grace and nature were not identical, nor did grace destroy nature. Grace transformed human nature by liberating it to attain to the union with God that it natively sought.

In the present ecological context, we can locate such high mysteries in a more preliminary stage: as the God of grace deals with human nature, so human nature should deal with the rest of nature. The human maintains its own uniqueness, in creativity and freedom: grace and nature are not the same. But the difference does not mean that the human can destroy nature, since it is meant to be a healing and preserving grace for it. But there is a transformative moment: grace, in this case the human, 'elevates', 'liberates' nature to a new fullness of communion and integrity.

Admittedly, contemporary theology is somewhat impatient of these venerable distinctions, preferring to reflect on the concreteness of human experience and history in the one focal, concrete mystery of Christ. I suspect, however, that these old distinctions are continuing to operate rather inarticulately in today's theological thought in the ecological and larger cosmic context. When nature is now seen, not primarily as a philosophical notion, nor as restricted to the domain of human sinfulness or creativity, but as a living, interrelated system of life and its manifold possibilities on this planet, a new set of correlations is called for.

It will not be a purely pragmatic relationship. That would leave too much unexamined. When the actor, the agent is left untroubled by larger demands for conversion, all solutions risk being merely the projection of stunted and unredeemed culture. A deeper participation in the process is required.

Nor is it merely a matter of a more ethical or newly ethical relationship. Without a vision of reality, without a vision of beauty, without the ongoing effort to understand the larger context and to make room for a more comprehensive sense of the common good in all its transcendent implications, ethics soon collapses into some form of oppressive moralism.

Nor, as we have said, is it simply a matter of a new aesthetic. Everything is not beautiful and life has piercing tragedies: the problem of evil is a problem.

On the other hand, an artistic model does add valuable suggestions of the transformative influence of the human within the whole natural order, particularly when that, in turn, is set within ultimate field of the Spirit. The resources and goal of such a transformation can reach their richest Christian expression only when set within the transforming experience of universal love.

7. The transformation of love

How is the contemplative dynamism of authentic faith located within the horizon of the ecological and cosmological relationships we have been exploring? Any answer to such a question would be incomplete without a specific consideration of Christian consciousness as informed by transcendent love.

By participating in the gift of love, the believer dwells in the universe as God's creation, to perceive it, however dimly or inchoatively, in its religious wholeness. Here contemplation blossoms, as it nurtures the ultimate relationality of our conscious being. The following points are worth consideration.

First, there is a kind of ultimate level of self-realization that mystical faith brings about. It is the realization of the self as radically communal. At the deepest roots of our consciousness, we are, in an essential sense, already involved with the ultimate mystery of our existence. Sebastian Moore has called it 'a pre-religious love affair with God', however unnamed or unknown such a God might be.[39] To be conscious at all is to be living a search for the ultimate affirmation of our relational being. Our consciousness unfolds within an horizon of ultimacy. What the ancients termed 'a natural desire to see God', or 'a desire for beatitude', modern thinkers express in more psychological terms. In both cases it is a matter of respecting the living all-involving thrust of our intelligence toward the meaning of all meaning, the sufficient reason for all our sufficient reasons, the worthwhileness of all our values, the ultimate affirmation of what living consciousness discloses us to be—meaning-making, truth-beholden, value-drawn beings. As Lonergan puts it in his lapidary manner:

> The question of God, then, lies within man's horizon. Man's

> transcendental subjectivity is mutilated or abolished, unless he is stretching forth toward the intelligible, the unconditioned, the good of value. The reach not of his attainment but of his intending is unrestricted. There lies within his horizon a region for the divine, a shrine of ultimate holiness. It cannot be ignored. The atheist may pronounce it empty. The agnostic may urge that he has found his investigation inconclusive. The contemporary humanist will refuse to allow the question to arise. But these negations presuppose the spark in our clod, our native orientation to the divine. [40]

Such a conclusion provokes a range of new questions. In what manner can this 'region for the divine' include all the regions of our living relationality? To what extent can this 'shrine of ultimate holiness' be the attainment of ultimate wholeness and universal belonging? To what degree can this 'native orientation to the divine' subsume the dynamics of the interconnective and emergent universe in which we are immersed? Christian faith finds the focus of its answer in the cosmic Christ: 'all things are made in him, through him and for him' (Col 1:16). But before concentrating on that focus, let us first be aware of the more connective quality of consciousness involved; and this leads to the next point.

Secondly, and this is in agreement with the emphasis of Teilhard mentioned previously, it is a matter of love: the *agape* of the New Testament, the *caritas* of the Catholic tradition. Though this is a distinctively Christian formulation, such love is understood to be a universal gift, however different religious traditions might express it. The original overture of human consciousness to the ultimate is kindled into intimacy through the self-giving mystery of the divine (Rm 5:5; 1 Jn 4:7–12). To participate in the transforming energies of divine love, in receiving and in giving it, is the crowning gift made to human existence: human consciousness homes to its final goal, to indwell a world radically transformed, to be freed from the void of absurdity as it is drawn into intimacy with the heart of the universe. As Lonergan puts it:

> As the question of God is implicit in all our questioning, so being in love with God is the basic fulfillment of our conscious intentionality. That fulfillment brings a deep-set joy that can remain despite humiliation, failure, privation, pain, betrayal, desertion. That fulfillment brings a radical peace, the peace the world cannot give. That fulfillment bears fruit in a love of one's neighbor that strives mightily to bring about the kingdom of God on earth. On the other hand, the absence of that fulfillment

> opens the way to the trivialisation of human life in the pursuit of fun, to the harshness of human life arising from the ruthless exercise of power, to despair about human welfare springing from the conviction that the universe is absurd.[41]

Here, too, questions arise. How does such a fulfillment of 'conscious intentionality'—the spontaneous thrust and overture of human consciousness in regard to the ultimate—begin now to embrace, not only the human neighbor, but the whole cosmos as a neighborhood? How does its energetic work for the sake of God's reign on earth stimulate the religious mind to include in its concerns the joy and peace of the whole of emerging process of creation? What has been said up to the present, and what remains still to be explored, is an attempt to answer such questions. But there is a third point to be emphasized.

Thirdly, then, such a self-realization in love is experienced as a conscious fulfillment. The experience of God is not immediately accessible through theological cultivation or philosophical refinement. It is something far more intimate and transformative, taking place in our actual living of relationships. Again, to quote Lonergan, in his analysis of the dynamics of consciousness:

> ... it is this consciousness as brought to fulfilment, as having undergone a conversion, as possessing a basis that may be broadened and deepened and heightened and enriched but not superseded, as ready to deliberate and judge and decide and act with the easy freedom of those who do good because they are in love. So the gift of God's love occupies the ground and root of the fourth and highest level of man's intentional consciousness. It takes over the peak of the soul, the *apex animae*.[42]

Such a fulfillment, with its outcome in a converted mind and heart, in the ultimate peace it intimates and in the freedom it inspires, begs to be properly set within the dynamisms and texture of contemporary culture. While it focuses on the ultimate, its basis can indeed be 'broadened and deepened, heightened and enriched' to include all else. In the current ecological crisis and in the context of a new cosmological sense of the universe, the challenge is to relate the ultimacy of the experience of God to all its forms of universality. It becomes a matter of integrating the ultimate in self-transcendence with the immanence of self-involvement. To live beyond this world and its cultural distortions holds multiple promise of living into it, with the special vitality of faith, hope and love. But another consideration arises.

Fourthly, this relationship of contemplative indwelling in the ultimate mystery is experienced as relational and dynamic. For it is manifest as a transition from isolation and emptiness to the connectedness and plenitude of communion:

> The transcendental notions, that is, our questions for intelligence, for reflection and for deliberation, constitute our capacity for self-transcendence. That capacity becomes an actuality when one falls in love. Then one's being becomes being-in-love. Such being-in-love has its antecedents, its causes, its conditions, its occasions. But once it has blossomed forth and as long as it lasts, it takes over. It is the first principle. From it flow one's desires and one's fears, one's joys and one's sorrows, one's discernment of values, one's decisions and one's deeds.[43]

It seems to me that this kind of radical affectivity is being freshly manifest in the integrative power of love coming into being in the ecological and cosmological context of today. For self-surrender to God, the 'first principle', is stimulating on the planet a new range of the desires, fears, joys and sorrows, discernment of values, decisions and activity as a loving collaboration with the God of creation. It is an expansion of loving God and of all in God, for:

> Being-in-love is of different kinds. There is the love of intimacy, of husband and wife, of parents and children. There is the love of one's fellow men with its fruit in the achievement of human welfare. There is the love of God with one's whole heart and soul, with all one's mind and with all one's strength (Mk 12:30). It is God's love flooding our hearts through the Holy Spirit given to us (Rom 5:5). It grounds the conviction of St Paul that 'there is nothing in death or life, in the realm of the spirits or superhuman powers, in the world as it is or the world as it shall be, in the forces of the universe, in heights or depths—nothing in all creation that can separate us from the love of God in Christ Jesus our Lord' (Rm 8:38f).[44]

It is now a matter of locating the transcendent, exclusive concentration of such love within the immanent, inclusive bearing of contemporary experience. As nothing in all creation can separate us from the loving source of our existence, nothing is more calculated to link us with that creation in a joyous participative companionship. Being in love with God attains a special contemporary radiance as a falling in love with the universe of divine creation, in all its variety and in all its generative groanings, there to find a new collaboration with the Creator Spirit.

8. Faith as contemplation

In such an understanding of the Christian experience of life, I would emphasize three features of the kind of consciousness that emerges.[45]

The first is a kind of immediate intimacy with the all-inclusive mystery. This is always in tension with the complex mediations of culture and religious tradition. These constitute an enormous and complex superstructure, articulated in word and ritual, institution and law, moral persuasion, all the variety of theologies and of scientific and philosophical theories, historical narrative, traditions, and worldviews. But through all these and reaching beyond them, faith connects with the inexpressible and ultimately inclusive realm of life and love. It leads to what Augustine termed 'a learned ignorance' that takes us beyond all formulations, even those most treasured in our traditions. It enjoys the presence of the *una quaedam summa res*, the one surpassing reality, for which there are no adequate words—for even the words 'reality' and 'cause' are limited by their earthbound significance. Mystics speak of inhaling the perfume of the divinity, of being wounded, inflamed, possessed by it, even while being unable to name it. The range of faith exceeds all other kinds of knowing. Rather matter-of-factly, St Thomas Aquinas asserts, 'the act of the believer does not terminate at a statement [about God] but at the Reality'.[46]

In tension with the various mediations of belief, of theology, and all other kinds of knowing, the mystical contact of faith cherishes its dark intimacy with the divine. A relevant point is that God is never to be fitted into a system as a factor or a process, but is the focus in which everything is centered, the realm of ultimate communion, the breath or atmosphere vitalizing all existence.

Secondly, I would add *interiority*. This is in tension with the ordinary apprehensions of a common-sense world in which God is thought of as 'up there', or as a big factor in the governing of the world, or a big 'thing' within it. Similarly such interiority contrasts with the intellectual world of theory, of theology or philosophy or scientific system. For the life of faith engages the whole being, from the depths of the heart, not just our minds or moral inclination or any other human capacity. In familiar biblical and spiritual discourse, God dwells in the heart of the mystic and the mystic dwells in God. A mutual indwelling, especially characterizing the writings of John. The mysticism of faith is not primarily busy in the articulation of a system or in making a synthesis. It lives in a presence, dwelling in it, surrendering to it. It is the eminent example of knowing by heart: the heart has its reasons which the head knows not of (Pascal). It is knowledge of familiarity, of the immersion of the whole self in the holy wholeness of things...[47]

Thirdly, the mystical orientation of faith is characterized by a *universal* and unrestricted scope. It lives from its own inherent excess. Here, too, the range of faith surpasses the particularities of doctrinal, theological or moral context. It breathes and 'groans' (Rm 8) in the mystery of the Holy Spirit who inspires hopes for a wholeness that cannot be thought, imagined or named. Such universality in the Spirit inspires the most radical form of catholicity as 'openness to the all' and as waiting for what is not yet...[48]

Hence, faith, at its most radical point, dwells in a totality which no thought or system or action can express or achieve. The experience of faith, in its immediate, interior fashion, is that of an inexpressible excess. Though such consciousness is marked by all the groaning limitations of obscurity, incompleteness and tension, it does not terminate in either absurdity or ultimate futility. Its universe is not a vast *reduction ad absurdum*. The basic character of its unfolding is that of surrender and adoration. It works more as a *reductio in mysterium* (Karl Rahner), as all things are brought back to the original and abiding mystery of everything and everyone.

In that sense of universal mystery, we bring this second circle of reflection to its conclusion.

1. See Alasdair McIntyre, *After Virtue*, Notre Dame Press, Notre Dame, Indiana, 1981.
2. For extensive treatment of this 'new story', see Thomas Berry, *The Dream of the Earth*, Sierra Club Books, San Francisco, 1988; and Brian Swimme, *The Universe is a Green Dragon: A Cosmic Creation Story*, Bear and Co., Santa Fe, 1984.
3. I owe this example originally to a Greenpeace publication. Unfortunately, I cannot verify the exact reference.
4. Thomas Berry, *The Dream of the Earth*, p. 134f.
5. As the basis for what follows, see Carl Sagan, *The Dragons of Eden: Speculations on the Evolution of Human Intelligence*, Random House, New York, 1977, pp. 14–16; and David S. Toolan, S.J., '"Nature is a Heraclitean Fire": Reflections on Cosmology in an Ecological Age', *Studies in the Spirituality of the Jesuits*, 23/5 November 1991, pp. 13–15.
6. Following the Advent antiphon, *Aperiatur terra et germinet salvatorem* ('Let the earth be opened and bud forth the saviour').
7. Thomas Aquinas, *Summa Theologica*, 1,1,6; 2–2, 45, 3.
8. ibid. 2–2, 45, 2.
9. Fritjof Capra and David Steindl-Rast with Thomas Matus, *Belonging to the Universe: Explorations on the Frontiers of Science and Spirituality*, Harper San Francisco, New York, 1991.
10. For abundant material and a great deal of dazzling insight on this inner journey as the rise of human consciousness, see Ken Wilber, *Up From Eden: A Transpersonal View of Human Evolution*, Shambhala, Boulder, 1983. For a stimulating collection of essays on some aspects of the new paradigm from a biological point of view, see William Irving Thompson (ed.), *Gaia: A Way of Knowing. Political Implications of the New Biology*, Lindesfarne, Press, Hudson, 1987.
11. B. Lonergan, *Method in Theology*, 'Method is not a set of rules to be followed

meticulously by a dolt. It is a framework of collaborative creativity' (p. xi).

12. For a fuller exposition of the New Paradigm, see Fritjof Capra, *The Tao of Physics*, (10th Edition), Flamingo, London, 1985; and *The Turning Point: Science, Society and the Rising Culture*, Flamingo, London, 1984.
13. Hence, in the collaborative method of theology, there are eight functional specialties, each with its own distinctive procedures. Since theological learning occurs in history, it is beholden both to the past and the present/future. Hence, on the empirical level, **research** examines the data of the past, and **communications** relates to the emerging context of the present. On the intellectual level, **interpretation** works to make sense of the past, and **systematics** is an attempt to create a holistic body of meaning for the present. On the rational level, **history** judges the achievement of the past, and **doctrines** states the priorities in the present. On the moral level, **dialectic** scrutinizes the conflicts of the past, and **foundations** seeks to objectify the most fruitful standpoint in the present. See B. Lonergan, *Method in Theology*, pp. 125–45.
14. See Richard Gelwick, *The Way of Discovery: An Introduction to the Thought of Michael Polanyi*, Oxford University Press, New York, 1977, pp. 135–6.
15. The original French title of the monumental work, Jacques Maritain, *The Degrees of Knowledge*, trans. Bernard Wall, Geoffrey Bles, London, 1937.
16. See Lateran IV (1215), in Neuner-Dupuis, *The Christian Faith*, Collins Liturgical Press, London, 1982, p. 109.
17. Here I am indebted to Mary Farrell Bednarowski, 'Literature of the New Age: A Review of Representative Sources', in *Religious Studies Review* 17/3, July 1991, pp. 209–16. The article is a judicious survey of representative views, as might be found in the productions of Marilyn Ferguson, *The Aquarian Conspiracy: Personal Growth and Social Transformation in the 1980s*, J. P. Tarcher, Los Angeles, 1980; David Spangler, *Emergence: The Rebirth of the Sacred*, Dell Publishing Co., New York, 1984; J. Gordon Melton, Jerome Clark and Aidan Kelly, *The New Age Encyclopedia*, Gale Research, Detroit, 1990; David Ray Griffin (ed.), *The Reenchantment of Science: Postmodern Proposals*, State University Press of NY, Albany, 1988; Louise B. Young, *The Unfinished Universe*, Simon and Schuster, New York, 1986; Brian Swimme, *The Universe is a Green Dragon: A Cosmic Creation Story*, Bear and Co., Santa Fe, 1984; John E. Lovelock, *Gaia: A New Look at Life on Earth*, OUP, Oxford, 1979; Bill Devall and George Sessions, *Deep Ecology: Living as if Nature Mattered*, Gibbs Smith, Salt Lake City, 1985; Ken Wilber, *No Boundary: Eastern and Western Approaches to Personal Growth*, Shambhala, Boston, 1979; Louise L. Hay, *You Can Heal Your Life*, Hay House, Santa Monica, CA, 1984; Starhawk (Miriam Simos), *Truth or Dare: Encounters with Power, Authority, and Mystery*, Harper and Row, San Francisco, 1987; Ram Dass and Paul Gorman, *How Can I Help? Stories and Reflections on Service*, Alfred A. Knopf, New York, 1988; David Toolan, *Facing West from California's Shores: A Jesuit's Journey into New Age Consciousness*, Crossroad, New York, 1987; and Matthew Fox, *The Coming of the Cosmic Christ: The Healing of Mother Earth and the Birth of a Global Renaissance*, Harper and Row, San Francisco, 1988. Of course, the possible list is endless—especially when even such works as this present exploration would be classified as 'New Age' by some hardy traditionalists!
18. As quoted in Jacques Maritain, *The Degrees of Knowledge*, p. ix.
19. For a specially valuable presentation of the Pauline diagnosis of the human condition, see Brendan Byrne, S.J., *Inheriting the Earth: The Pauline Basis for a Spirituality for Our Time*, St Paul Publications, Homebush, NSW, 1990. See, too, this author's remarks on the Pauline 'New Age' compared with the current phenomenon, pp. 12; 24.
20. See R. Sheldrake, *The Rebirth of Nature*, Bantam Books, London,1992, pp. 49f.
21. For a critique of the totalitarian attitudes of science, see Bryan Appleyard, *Understanding the Present: Science and the Soul of Modern Man*, Picador, London, 1992; and Mary Midgley, *Science as Salvation: A Modern Myth and its Meaning*, Routledge, London, 1992.
22. An outstanding resource here is Thomas M. King, S.J., *Teilhard de Chardin*,

Michael Glazier, Wilmington, Delaware, 1988. See especially pp. 13–64.

23. Pierre Teilhard de Chardin, *Towards the Future*, trans. Rene Hague, Harcourt Brace Jovanovich, New York, 1975, pp. 209f.
24. For further details, see Thomas M. King, *Teilhard de Chardin*, pp. 15–19. Here Teilhard is more akin to Buber and Zaehner than to Huxley and Stace. More profoundly, his emphasis on particularity represents a Western philosophical and mystical tradition in contrast to the classic Eastern types. But that is a matter requiring much deeper exploration than is possible here.
25. See my *Trinity of Love: A Theology of the Christian God*, Michael Glazier, Wilmington, Delaware, 1988, especially pp. 234–48 for some suggestions toward a trinitarian model for the resolution of these divergences.
26. David Toolan, *Facing West from California's Shores: A Jesuit's Journey into New Age Consciousness*, op.cit., p. 30.
27. How a healthy 'deconstruction' is at work in the emerging eco-cosmological context has yet to be worked out. In beginning to address the problem, I have found the following valuable references: Kevin Hart, *The Trespass of the Sign: Deconstruction, Theology and Philosophy*, Cambridge University Press, Melbourne, 1989; and Ronald H. McKinney, 'Toward the Resolution of the Paradigm Conflict: Holism versus Postmodernism', in *Philosophy Today*, Winter 1988, pp. 299–311.
28. Pope John Paul II, *Peace with God the Creator: Peace with all Creation*, St Paul Publications, Homebush, NSW, 1990, #13, para. 3.
29. A work of enduring value is William F. Lynch, *Images of Hope: Imagination as Healer of the Hopeless*, Notre Dame, University of Notre Dame Press, 1965.
30. Brendan Lovett, *Life Before Death: Inculturating Hope*, Claretian Press, Quezon City, 198?, p. 7. For a practical way of responding see Albert LaChance, *Greenspirit: Twelve Steps in Ecological Spirituality*, Element, Rockport, Massachusetts, 1991.
31. The importance, and the depth of analysis involved, of this 'psychic conversion' in a world-cultural context is explored in the important book, Robert Doran, *Theology and the Dialectics of History*, Univesity of Toronto Press, Toronto, 1990.
32. Pope John Paul II, *Peace with God the Creator: Peace with all Creation*, #13, par. 1–2.
33. Douglas M. Meeks, *God the Economist: The Doctrine of God and the Political Economy*, Fortress Press, Minneapolis, 1989, p. 34.
34. For a rich context of discussion, see Neil Ormerod, 'Renewing the Earth—Renewing Theology', *Pacifica* 4, 1991, pp. 295–306; and Denis Edwards, 'The Integrity of Creation: Catholic Social Teaching for an Ecological Age'. *Pacifica* 5, 1992, pp. 182–203.
35. For excellent documentation and reflection on this issue, see Denis Edwards, 'The Integrity of Creation: Catholic Social Teaching for an Ecological Age', *Pacifica* 5(1992), pp. 182–203.
36. Neil Ormerod, 'Renewing the Earth—Renewing Theology', *Pacifica* 4 1991, pp. 295–306; and Robert Faricy, *Wind and Sea Obey Him: Approaches to a Theology of Nature*. SCM, London, 1982, pp. 53–61.
37. On the more general aesthetic consideration, Pope John Paul II makes an interesting comment:
    The aesthetic value of creation cannot be overlooked. Our very contact with nature has a deep restorative power; the contemplation of its magnificence imparts peace and serenity. The Bible speaks again and again of the goodness and beauty of creation which is called on to glorify God (Gn 1:4ff; Ps 8:2; 104:1ff; Ws 13:3–5; Si 39:16–33; 43:1) *Peace with God the Creator: Peace with all Creation* #14, n.1
38. This point awaits further application in the later chapter on 'Wholesome Sex'.
39. See Sebastian Moore, *The Fire and the Rose are One*, Darton, Longman and Todd, London, 1980, pp. 5–28.
40. B. Lonergan, *Method in Theology*, p. 103. For further applications of Lonergan's work to spirituality see John Bathersby, *The Foundations of Christian Spirituality in Bernard Lonergan, S.J.*, N. Pecheux, Rome, 1982; and the regrettably unpub-

lished dissertation of Frank Fletcher, *Exploring Christian Theology's Foundations in Religious Experience*, Melbourne College of Divinity, 1982.

41. Bernard Lonergan, *Method in Theology*, p. 105.
42. Ibid., p. 107.
43. Ibid., p. 105.
44. Ibid., p. 105.
45. These points beg to be developed in a larger context as, to give one excellent example, it is described in Gabriel Gomes, *Song of the Skylark I-II: Foundations of Experiential Religion*, University of America Press, Lanham, Maryland, 1991.
46. Thomas Aquinas, *Summa Theologica*, 2–2, 1, 2 and 2. For relevant material, see G. Gomes, *The Song of the Skylark I*, pp. 211–32.
47. See Gomes, *The Song of the Skylark I*, pp. 216–23; 239; 295.
48. Ibid., pp. 233–48.

SECTION III

A Third Circle of Connections: The Logos in the Cosmos

In this next circle of connections, I wish to emphasize the distinctive Christian and theological elements in the spiral of our reflections. First of all, then, a note on the Word incarnate, the Logos uttered into the cosmos of our experience: 'For God so loved the world[1] that he gave his only Son... Indeed, God did not send his Son into the world to condemn the world, but in order that the world might be saved through him.' (Jn 3:16–17).

1. Experiencing the Word

At first glance, given the cosmic tone of its primary sources, especially in the Pauline and Johannine writings, it would appear strange if Christian faith were deaf to such resonances. Admittedly, the remarkable biblical renewal this century has in fact not been notably interested in interpreting the biblical documents in the context that concerns us here. You can't do everything all at once; not even biblical scholars. Their main challenge, met with enormous erudition, was to establish not the cosmic, but the historical connection of the Christian faith. 'The Historical Jesus' has been, up to the present, the object of meticulous examination of the biblical documents. This has been further extended into a deeper literary appreciation of the narrative quality of Christian sources.

It may sound ungracious to suggest that the very skills and expertise shown in this area of Christian investigation have tended to close it off from more cosmic considerations.[2] But, as I say, no one can do everything all at once. Moreover, both the historical connection and narrative form put us in the position to give a new hearing to the larger story of creation in more critically real terms. The critical historical placement of 'the Jesus Event' can only bring an added realism into the consideration of his significance at this present juncture of the history of Christian faith in its efforts to engage a world immeasurably more vast and intricate than the ancients could have imagined.

The scope of the Christian horizon can be quickly suggested by referring to some key passages from the prologue of John's Gospel:[3]

> In the beginning was the Word, and the Word was with God, and the Word was God. He was in the beginning with God. All things came into being through him, and without him not one thing came into being. (Jn 1:1–3)

Such a passage brims with promise as we read it in the current context. How might this *Logos* embrace in its meaning all the other *logoi* (meanings) of our human explorations—ecology, cosmology, and all the *dialogoi* that result? For John, it seems clear enough that there is an inexhaustible original relevance of the Word. The divine self-utterance is all-creative: 'all things' came into existence through him, and depend on him in their origins; all meaning is a resonance of this original divine meaning.

Indeed, in the Johannine context, 'the Word' conjures up numerous associations to a variety of other biblical and philosophical understandings of 'word' (*logos*, *dabhar*). For instance, it was a pervasive term in Greek philosophy, and Heraclitus of Ephesus (where John possibly wrote the Gospel) introduced it as a philosophical principle six centuries beforehand. In later Stoic philosophy the Logos figured as a quasi-divine cosmic principle of order. Philo, bringing together Greek and Jewish thought, used the term over a thousand times in his writings to designate a kind of created intermediary between God and the world: the divine image is mirrored in the coherence of the cosmos and in the human soul itself. Later Gnostic thinking stressed more the divinity and immateriality of the Word.

No doubt these were influences and associations—how could they be avoided? Nonetheless, it is now agreed that the dominant influence is to be found in Jewish theology itself, where 'the divine word' was understood in more active and concrete terms, intimately connected to Spirit or Breath of God, present within history. For example, like the divine utterance of Genesis, John's Word is 'in the beginning'; it is all-creative; it overcomes the darkness of chaos, and gives life. Perhaps most of all, the strongest links are to the personified 'wisdom' of the Sapiential literature. This concrete, active emphasis culminates in the prologue itself as the Word becomes 'flesh to dwell amongst us'.

Later Christian theology was not slow to exploit a whole range of analogical references to the Word. For example, by linking the generation of the Son to the utterance of the Word, early Christian thought was enabled to repel the charge made by cultured adversaries that it was a regression to the naively mythological. The divine generation happened, not by physical generation, but by a spiritual conception or emanation. Similarly, it served in relating the eternal genesis of the Word/Son within the divine realm to the genesis of Jesus the Incarnate Word and Son within the world and time. Most importantly, the Word became a principle of Christian conversation with a number of forms of the human culture: the Word that has been finally revealed in Christ has previously enlightened both the prophets of Israel and the sages of antiquity. And in current understandings of the Church's mission to the world, the notion of transcendent incarnate Word underpins all the forms of 'dialogue', the unending challenge of listening to the Word through the many words of human science, scholarship and religious tradition.

Though the notion or symbol of the Word is of such fundamental biblical importance, it is striking how often it is linked to the symbol of light.[4] The life of faith in Christ is expressed as an illumination that comes through receptive listening. The radiance of the Word places believers within a divinely wrought universe. Human existence becomes translucent as an experience of the Word-shaped world. As the Word is definitive and irrevocable in its meaning, it is a light that cannot be extinguished:

> What has come into being in him was life, and the life was the light of all people. The light shines in the darkness, and the darkness did not overcome it (Jn 1:4–5).

As faith approaches the mystery of the universe created in the Word, it comprehends all history as a progressive, universal illumination. We are all involved in a great luminous field of meaning:

> The true light, which enlightens everyone, was coming into the world. (v.9)

This process culminates at the point where the Word enters the human conversation as a distinctive presence, to be embodied in the communicative reality of our world:

> And the Word became flesh and lived among us. (v.14)

The originality of Christian faith is underscored in Augustine's remark: he had found the equivalents of most Christian doctrines in pagan authors, but what he had never found was that the Word became 'flesh'.[5] Life is transformed in the light of ultimate meaning; human consciousness now expands in a new luminous horizon; the Word has become a participant in the human conversation. The self-expression of God is from this point irreversibly linked not only to the transcendent capacities of the human spirit, but in becoming human is immersed in the physical, chemical, biological, psychological and cultural dynamics of human emergence—'flesh'. The identification of the Word with the realities of created existence must account, in some measure, for Christ's self-identification as 'living bread' (Jn 6:51ff), as 'the door of the sheepfold' (Jn 10:7), as 'the good shepherd' (10:11), as 'the true vine' (Jn 15:1). His use of the earthly symbols of eating and drinking, of sowing and harvesting, of fishing and tending flocks, of water, light, wind and flame, tend to present nature as a great parable of Incarnation.

The plenitude (*pleroma*) of divine self-utterance into the world is the decisive mutation in human history: 'From his fullness we have all received, grace upon grace' (v.16).

Despite such intense expressions of Christian experience as addressed, illumined, enlivened and accompanied by the Word, the ultimate mystery is not diminished or contained by human intelligence. The Word incarnate mediates the inexpressible 'more' of the Father. The silence out of which the Word is spoken, the radiance out of which this light has shone, remains. God is the abiding mystery, the all-encompassing reality of that love which, beyond the categories of this world, welcomes creatures into 'his house of many rooms' (Jn 14:2).

The distinctive chiaroscuro of Christian experience is expressed in these words of the Word: 'No one has ever seen God. It is God the only Son, who is close to the Father's heart, who has made him known' (v.18). If John's Gospel begins with the shining promise of the Word dwelling amongst us in the flesh, it concludes by supposing 'that the world itself could not contain the books that would be written'(Jn 21:25) in any attempt to grasp the full significance of what has happened.

The Johannine prologue is not unlike a great classic poem, demanding ever deeper reflection, yet inexhaustible in its potential range of reference. It draws the believer into a luminous space within an horizon of limitless exploration. Within that space, Christian history has to keep trying to 'word the Word' in ways that illumine the darkness of the successive crises of history. As the Gospel unfolds, the Word, by becoming flesh, becomes a story, a conversation, a question, a theology, a prayer and a promise—eventually a silence, in death and in resurrection, before which all words must fail. But there are pauses in the conversation when the believing mind can recollect itself and pose new questions. How does such light illumine our planetary coexistence? How can such a culminating instance of God 'so loving the cosmos' find new expression within the astonishing dimensions of the cosmos of our knowledge today?

2. Wording the Word

It would be rather fundamentalist to expect to find in the New Testament instructions on modern ecological concerns or on current cosmological theory. For that reason, I have emphasized more the horizon, or the background radiation, you might say, in which such matters might be assimilated into Christian awareness. This can be made more specific by noting the movement of Christian response to the Word, as it is instanced in three major arcs of meaning.[6] The New Testament documents communicate not the conclusions we might reach in this much later age, but open-ended ways of thinking and speaking suitable for integrating the range of each age's experience into the mystery of Christ.

The three major arcs of meaning I refer to can be termed: the rhetoric of fulfillment; the rhetoric of participation; and the rhetoric of cosmic extension. I use the word 'rhetoric' not, of course, as something akin to 'mere rhetoric', but in a more classical sense: the creative effort to bring experience to expression, to word reality ; in this case, to seek for words, for ways of speaking, ever more worthy of the Word.

The first rhetoric, then, deals with fulfillment: the Word incarnate fulfills the prayers and promises, the hopes and anticipations of what preceded him: 'The law was given through Moses; grace and truth came through Jesus Christ' (v.17). In Christ 'we have obtained an inheritance, having been destined according to the purpose of him who accomplishes all things according to his counsel and will' (Ep 1:11). For 'long ago God spoke to our ancestors in many and various ways by the prophets, but in these last days, he has spoken to us by a Son' (Heb 1:1f). This kind of fulfillment is present primarily in the Scriptures as a fulfillment in history, above all that of the hopes of Israel awaiting its messianic deliverance; and beyond that, the hope that the good creation of God's original intention would reach fulfillment in the New Adam.

Later patristic theologies, such as those of Irenaeus of Lyons and Origen, would further explore the all-fulfilling role of Christ. The later Scholastic theology of the Middle Ages, with its notion of Christ as the perfect divine image, would set such a moment of fulfilment in a metaphysical universe. However, even for the great Scholastics, the movement of history took place within a universe of fixed natures, a vertically arranged 'chain of being', so to speak. Today the notion of fixed universe has changed dramatically to an evolving, emerging cosmic process: the chain of being is horizontally linked, holding matter, consciousness, spirit together as ever more complex phases of universal becoming. This new understanding of the universal process has enabled theology to present the mystery of Jesus Christ less in terms of the Word 'coming down from heaven', or of God 'sending his Son' into a fixed world of creation. The emphasis now is on God as the creative force enabling the continuous self-transcending process of the world, until it finally reaches the point of being able to receive the fullness of its original mystery: in the consciousness of Christ, we have realized, from different points of view, creation accepting God as the ultimate dimension of its life; and God accepting creation as personally his own. Hence, Christ represents a decisive mutation offered to human consciousness. A new creation, life to the full, a transformation of all things in Christ has begun.

The second manner of wording the Word incarnate is that of participation. It deals with our common connection in the reality of Christ: 'in him was life' (v.4) and '...from his fullness we have received...' (v. 16). He is the vine, we the branches (Jn 15: 5f; cf. 14:6; Mt 11:27; Ac 4:12). The universal is made available to us in the particular: 'I am the way, the truth and the life. No one comes to the Father except through me' (Jn 14:6).

Contemporary theology explores our common participation in God's self-communication in Christ in new ways. Teilhard, for example, sees Christ as the finality of creation already present. The whole meaning of the process of the world's continuing development is to incorporate all into the incarnate mystery of Christ. Every moment of time is impregnated with his presence. The whole cosmic process is Christ-bearing: 'The prodigious expanses of time that preceded the first Christmas were not empty of Christ, for they were imbued with his power.'[7] Material creation is not left behind, for it too finds its destiny by being incorporated into his transfigured Body: the whole of creation, physical and spiritual, is like a eucharistic host, offering itself to be consecrated, and thereby transformed into the Risen Lord.

The third arc of wording the Word is that of cosmic extension. It elaborates the meaning of the mystery of Christ as one of universal inclusiveness: 'all things came into being through him, and without him not one thing came into being' (Jn 1:3; cf. Gn 1:1; Pr 8: 27–30; Heb 1:2). He is the focal element in God's 'so loving the world'. In him is inscribed the 'whole story', for it is God's intention 'to gather all things up in him, things in heaven and things on earth' (Ep 1:10), just as he is 'before all things, and in him all things hold together' (Col 1:17).

This kind of biblical expression inspires all the variety of modern spiritualities and theologies of 'the Cosmic Christ'.[8] The wider the extension and the deeper the inclusion of the 'all' of present experience, the richer the apprehension of the Christian mystery. Conversely, the more intimate our union with Christ, the more 'the all' becomes a universe of grace. Admittedly, the challenge today is to integrate into the intimacy and universality of faith not only the fifteen billion years of the world's emergence, but also a capacity to cherish as God's creation, the varied commonwealth of life in which we share. Whatever there is yet to occur in the world of our experience, whatever dimensions or dynamics there are in the cosmos of our present and further exploration, Christian faith must find its focus in him who is the consistency and coherence, the firstborn and the end of all creation. As Teilhard reminds us:

> Christ must be kept as large as creation and remains its Head. No matter how large we discover the world to be, the figure of Jesus, risen from the dead, must embrace it in its entirety.[9]

Such, then, is a small indication of three fundamental arcs of Christian rhetoric. Let me stress, they indicate fundamental 'ways of thinking', and remain so. This is not to say the creativity of their respective trajectories was always fully respected in later doctrinal councils and theologies. Such articulations were limited to particular contexts where creativity had to yield to other responsibilities. It should not be forgotten, however, that even the most austere doctrinal definition can only be understood as the more or less successful effort to objectify some dimension of Christian experience that was being endangered. The Christian Fact resides in the whole life of the Church. Hence, doctrinal definitions were always understood to emerge out of that life and to return to it, to contribute the distinctive enthusiasm that can only come from intellectual clarity. The liberation of the heart and the confident momentum of life cannot long endure without the objectifying, clarifying, position-taking activity of mind.

But whether we speak of the creative rhetoric of faith, or the more precise articulations of belief regarding the incarnation of the Word in our world, Christian thought is now confronted with an enormous challenge and opportunity. As the next millennium unfolds, how can Christianity retrieve and rework its most basic doctrine in the light of the awe-inspiring dimensions of the world as it is now coming to be known? How can the process of God's coming-to-be in the world be meshed with the manifold processes of the world's emergence? This will remain one of the most energizing points for Christian thinking in the new era before us.[10]

3. Incarnation: from answer to question

Some dimensions of the challenge and of the resources with which it can be addressed can be suggested by delaying briefly on the classic doctrine of the Council of Chalcedon in 451. I will try to indicate what it achieved, and the questions it now faces us with. The relevant section is as follows:

> One and the same Son, Our Lord Jesus Christ, the same perfect in divinity, perfect in humanity, the same truly God, and truly man with a rational soul and body, consubstantial with the Father according to his divinity, consubstantial with us in his humanity, like us in all things except sin, begotten of the Father before the ages in his divinity, but in these last days, for us and our salvation, born of the Virgin Mary, Mother of God, in his humanity, in two natures without confusion or change, without division or separation, the distinction of the natures not being abolished by the union, but on the contrary, the properties of each nature remaining intact, and coming together in the one person or hypostasis, not by being split or divided into two persons but one only unique Son, God, Word, Lord Jesus Christ, as formerly the prophets spoke of him, as Our Lord Jesus Christ himself instructed us, as the symbol of the Fathers has handed down to us.

In this complex grammar of the meaningfulness of Incarnation, there are certain things to be noted:

First, the concentration of faith bears not on a theory, nor on a theology but on the singular, concrete reality of the person, 'One and the same, Our Lord Jesus Christ'. The 'he' of the Gospels, the 'you' addressed in prayer, the 'I' that summons forth faith, is the all-meaningful datum. Christian consciousness is focused in a living relationship with him, just as the 'he' is only understood in relationship to the One from whom he came, and to the world for which he came. The focus of faith is on the concreteness of the relationships that Jesus embodied. It places us in a cosmos of ultimate relationships and mutual presence.

Secondly, the elaboration of this realm of relationship is made in a quasi-philosophical language. The more figurative and rhetorical scriptural expressions are transcended in the interests of affirming the reality of Christ to meet with the demands of philosophically differentiated culture. The task of precisely denoting the objective form of the Christian fact precludes a comprehension of the whole range of tacit connotation. Hence, we are presented with an austere duality of divine and human 'natures'. Nonetheless, 'nature' is employed as an open-ended kind of notion—to mean whatever the divine or the human are, or can be. There is little effort to define either, except to this extent:

With regard to God, the dogma recognizes that there is a generative process, a vital genesis, within God—Christ is Son, Word, begotten of the Father—and that this divine realm of genesis, though outside of time ('before the ages') decisively modifies the meaning and quality of time, to condense it into the finality 'of these last days'. With regard to the humanity concerned, Chalcedon stresses the realism of Christ's human being (with a rational soul and body, and of human birth at a particular time): the divine Word is a distinct presence in the world—embodied in its processes, consciously communicating within it, brought forth by it, from the womb of a human mother.

Thirdly, the focus of the definition finds its special intensity in affirming that Jesus Christ is truly what God is (consubstantial with the Father) and truly what human beings are (consubstantial with us, like us in all things except sin). Within the horizon of human experience, the mystery of God and the reality of the human have come together in a unique, irreversible manner—later to be called the 'hypostatic union'. God has communicated the God-self to the human, to be self-involved in its ultimate genesis.

Fourthly, the resultant identity of Jesus Christ as God's Son and a member of the human race does not confuse the duality of the natures. Communion, not dissolution, is the goal. Humanity is not changed into something else, neither is the divinity reduced to some anthropomorphic form ('... the distinction of the natures nor being abolished by the union ... the properties of each nature remaining intact'). Jesus is not a fusion of the divine and the human, neither a demigod nor a kind of centaur, nor two persons, one human and one divine. The divine and the human, the creator and the creature, are left free to be themselves. In the unity of his identity, Jesus is both God and a man—God uniquely given into the human world, and the human uniquely identified with God the originator of the world. All this 'without confusion or change, without division or separation'. In him, the world is drawn into union with its original mystery. In him, that original mystery becomes part of the history of the world.

Fifthly, the decisive phrase 'for us and our salvation'. The rather objective, philosophical terms of the definition are brought back to their primary focus—the appreciation of how God is radically 'for us', not in terms of theological information or metaphysical system, but as the divine self-engagement with human beings in their search for true life.

The four and a half centuries of Christian reflection expressed in such a statement occupies a classical position in Christian thought. But times change and other worldviews develop. It is customary today to stress, not only the inadequacy of the philosophical terminology involved (the meanings of 'person' and 'nature' have changed through the centuries to have almost opposite meanings in contrast to the Conciliar usage), but the enormous change in the basic apprehension of reality itself. Here the following points can profitably be made to elucidate the modern context:[11]

First, there is the basic problem of the notion of God itself. Former cultural times, right up to the modern era, were theistic. A notion of the divine was part of the worldview, either in its popular or sophisticated forms: God was the transcendent beginning and end of all things. A generalized religious, intellectual, and moral conception of a divine absolute was a presupposition in any reasonable discourse. Ironically, Christianity, in its effort to articulate its own distinctive trinitarian apprehension of God, had to struggle against congealed forms of such theistic notions; and no doubt still has to, to the degree they survive.

Today that theistic presupposition is culturally absent.[12] Contemporary culture, profoundly affected by the great atheistic masters of suspicion—Darwin, Marx, Nietzsche, Freud—has unfolded in a religious void. Thus, the previously culturally legitimated concept of the divine has yielded to a culturally legitimated denial of God. As that religious void loomed in modern consciousness as a world without meaning and purpose, profoundly indifferent to human achievement or moral value, responses have been many and various. Fundamentalism thrives, New Age spiritualities emerge, theologies of liberation or mysticism criticize the idolatries of the past, and find new opportunities for affirming the lost sense of shared mystery either in solidarity with the poor, or in the cultivation of contemplative experience.

Most strikingly, the question of God has emerged in science itself. What is the ultimate reality intimated in the vast self-organizing cosmic process in which we are involved? Whatever the case, the notion of 'God' has ceased to be present as a culturally available answer. It has moved into the form of a radical kind of questioning. What kind of God might be acknowledged in a cosmos conscious of itself in human minds and hearts, in all its chaos and entropy, yet in all its connections, self-organization and novelty? How, in such a context, then, is Christ the revelation of God?

Secondly, there is a basic problem with the notion of a human nature. In previous times, the human was defined, following Aristotle, as *zoon logikon*, *animal rationale*: there is a genus of the animal that has the specific difference of the capacity to use reflective thought and to communicate in speech. The elemental richness of this definition of the human, as in the Chalcedonian doctrine, was usually reduced to the requirement that the nature of the human demanded a body and a soul, the material component and the spiritual. Even if Aristotle and Aquinas, in different ways, overcame the Platonic tendency of seeing the soul more or less imprisoned in matter, popular conceptions were as a rule dualistic.

Today such dualism is either denied or rendered questionable by a more concrete empirical exploration of the human: here the human is not a fixed nature, but that which is examined in human consciousness (psychology); that which articulates itself in social and cultural forms (anthropology); that which emerges at the end of a long evolutionary process (biology); that which enables the universe to become conscious of itself (cosmology); that which is forming itself in freedom (history). Contemporary reflection on the human yields no simple definition of the human, but rather examines all the varied data in different and often merely juxtaposed contexts.

The traditional philosophical definition of the human as *animal rationale*, and the consequent popular schematic description of the human as a composite of body and soul, have obviously been displaced by a larger problematic. Given our historical experience of the problem of evil (in all its ecological, political, economic dimensions), the meaning of the human is not so much an answer but a question: what are we to become? The older classical culture, with its precise philosophical definition of what it was to be human, has now yielded to an empirical notion of culture: what are we to make of ourselves, globally, in the exchanges of many cultures, with the corresponding ways of being human they represent? Given the manifold potential of human freedom, what should we decide to be? What are the limits to such self-determination? What values should sustain it? What might we hope for? What do we ultimately share? How, finally, is Christ 'consubstantial with us' in such a humanity?

Thirdly, a question emerges regarding the connection between the two 'natures' in which the person of Christ subsists. Where before, Chalcedon could simply state the basic Christian fact in terms of Jesus Christ being one person in two natures, the divine and the human, modern reflection has concentrated on the reasons for this mysterious union. What is there about the divine reality that inclines it to express and utter itself in the human through the incarnation? Why should the creator of all enter creation in *this* way? Is the advent of the Word to the world an arbitrary decision of the divine? Or, is the deeper possibility to be considered, that the divine vitality is ever one of self-communication, ever intent on uttering itself into what is other; indeed, involved in creating a universe so that 'God may be all in all' (1 Co 15:28), in a realm of ultimate life and communion?

Fourthly, there is the related question dealing with human existence itself. What is there in human existence that can greet the event of God's becoming human, not in dread as something destructive of our humanity, not in mere wonderment as a divine visitation, but as a transforming event, a mutation in the genesis of the human—a grace healing, perfecting and transforming human nature, and the whole cosmic body of our coexistence?

Fifthly, the universe of the Chalcedonian definition is a very limited one. Its worldview is one of fixed, hierarchical arrangement. The universe is thus characterized by a certain ordered fixity. It is a cosmos of definable entities and intelligible order. It is structured in a gradation of natures and capacities. It knows its problems and knows its needs—the chief of which is salvation and union with the absolute beyond all the fragmentation and flux of time: the One who is above comes to draw humankind into his own immortal and eternal sphere. The destiny of the human is to be divinized.

In contrast to such a worldview is that of contemporary science and culture. Hitherto unimaginable dimensions of time and space and energy have entered into our minds today. We understand and increasingly experience this world as one of amazing, intricate and even chaotic emergence. It is a process, unfolding through billions of years in increasing differentiation of physical, chemical and biological forms. It expresses itself in different and successive levels of organization, from the quantum behavior of particles to the formation of increasingly complex molecules to the emergence of life, from the protozoa to what we can now call plant, animal and human. Such emergent differentiation and interiority, heading to the phenomenon of the human, dawns as a universe, a vast communion, a cosmic emergence already of unimaginable duration, begging for ever more creative meanings to interpret it.

How does the incarnation fit into such a scheme? How, why does God become human in such a cosmic unfolding? How is Christ incarnate, not only in an individual human body, but in the cosmic body of emergent reality? These are the larger questions brought dramatically into consideration by the writings of Teilhard de Chardin.[13] The divine appears more now as the limitless matrix of life out of which the whole process has emerged, to communicate itself in a personal way at the beginning of this 'second day', in terms of our earlier illustration, in order to offer a new and final integration of the all in a common destiny.

And so, from the above five points of view, the great doctrinal inheritance of Chalcedon explodes into whole ranges of new questions. Perhaps they are best when most simply put: what if the incarnation were true? What difference would it make? What taste for life would it offer? What sense of the whole would it confer?

To anyone already a committed Christian, such questions seem to entertain a needless doubt. They seem to imply a deeper capitulation to a relativistic and agnostic mood, perhaps in the hope of evoking a little bit more consideration from well-meaning non-believers. But to leave the matter there is to place Christian faith in an endless effort of self-justification in a world of disbelief.

On the other hand, such a question can profitably stir a more creative response from Christians themselves. Given the complexity of human consciousness and the fragmented character of modern culture, it is possible to live 'as if the incarnation were not true' in more or less vast areas of human experience. The consequences of the original, radical Christian affirmation remain untapped. The incarnation is left in doctrinal isolation, cut off from a larger imagining. It is a completed viewpoint rather than a developing horizon, leaving the believer content with doctrinal certitude rather than stirred to wonder. The dogma congeals into a statement, instead of inspiring new questions.

Without losing confidence in its basic affirmation, living much closer to the questionable, questing, wondering energies of faith can mean real gains for the vitality of Christian imagination.

So, what if it were true that in Jesus of Nazareth the divine has expressed itself? What difference would it make in the whole world of our experience? What if it were true that the Word, the self-expression of limitless originating mystery, had become flesh, uttered itself into what seems most distant from any notion of the divine? What if it were true that the divine has uttered itself into the matter and processes, into the emergence and into the community of this planetary life? What if it were true that the God, from whom the universe has come and is held in being, has become with us, an earthling, to emerge out of the fertility of this process of life? What if it were true that the universe itself had become newly conscious of itself in the mind and heart of the Word made flesh, the Word incarnate, en-worlded, encosmified? What if it were true that the procession of the Word from the Father in the Spirit—not 'any kind of word, but a Word breathing love' as Aquinas would say[14]—was the deepest structural dynamism of the world process itself?

What if this were the way reality is? What difference does it make to our sense of God, of ourselves, of the world, of the universe, in its origins and in its future?

4. The Word and the worlds of meaning

I am not hereby implying that there are only questions. Indeed, new answers are being shaped not only in theology, but in new doctrinal expressions of faith. One convenient source for documenting this is Vatican II's *Gaudium et Spes* as it gives expression to the cosmic, historical, social and cultural understanding of human existence. One notes the change of emphasis, along with a certain overture to a less philosophically burdened style of thinking—in some ways a reversion to an earlier style of Christian thought characteristic of Irenaeus of Lyons and Origen, and yet owing much to the influence of Teilhard and the evolutionary perspective of Rahner.

Before commenting on some conciliar texts, I feel it is important to take note of the meaning-making process that is in evidence in the creativity of the Church's expression.

Meaning itself is a strange 'in between' notion. In 'meaning' something, we certainly intend to make a true statement. We are attempting to deal with reality. But that reality is disclosed only within an horizon lit by many kinds of meaning-full acts—and even then, in its deepest reaches and extent, existence remains an all-inclusive mystery, inciting us to meaning, but eluding any complete comprehension. To express the matter another way, it is not as though we simply put a meaning on things already somehow 'there'. Rather, through our efforts to be meaning-making, we find our immersion in reality illuminated in limitless ways. Without it, the world would be meaning-less. We would have no name, no language, no history, no laws, no thought, no art, and certainly no religion. Even to ask a question implies our involvement in a meaningful world.

To be suddenly left stranded in our worlds of ordinary meaning, say, when we are touched by dreadful suffering, or pierced with the absurdity of the situation, or shocked by a sudden death, or drawn into the divine presence, is profoundly disconcerting. For being human is a continuing effort to live in a meaning-full world, a world shot-through with significance and intelligibility.

Without going into all the complexity of this topic, we usually associate meaning simply with what is meant in the world of known objects. I mean 'kiwis', not 'kangaroos'; I mean that the cows are in the meadow, not that they jumped over the moon; I mean that the moon is made of rock, not of cheese. Hence, the most noticeable function of meaning is that it orients us in an objective world. It is *cognitive*, as philosophers would say. This world of objects is one of limitless scope, from quarks to quasars, from photons to the world of faith, from protozoa to the Trinity itself. Early in human development we learn to distinguish between real and imaginary objects, eventually to arrive at distinctions between astrology and astronomy, between myth and history, between truth and fantasy, between well-founded belief and childish credulity. In that apparently sturdy world of cognitive meaning, we are constantly encouraged 'to say what you mean and to mean what you say'. It is a world in which knowing supposes an increasing ability to name and to refer to a world of objects, in their distinctiveness and connections.

Now this has been very much the world of Christian faith with its hardy intellectual inheritance from Greek philosophy. Thus, you have the articles of the creed and the various church doctrines that teach that this is the case, not that. God's Word is, thus, the supreme object which has entered into the flesh of our world of objects. Theological systems characteristically endeavor to bring these doctrines into a meaningful whole, and to relate that world of the objects of faith to the world of science, philosophy, and the human sciences in general. Manifestly, this remains a great challenge at the moment: to take aboard the meaning-system of Christian faith all the new things and relationships that ecology or cosmology are increasingly discovering.

As an example of the greater cognitive range of meaning, we can cite the following passage:

> The Word of God, through whom all things were made, was made flesh so that as perfect man he could save all human beings and sum up all things in himself. The Lord is the goal of human history, the focal point of the desires of history and civilization, the center of humankind, the joy of all hearts, the fulfillment of all aspirations. It is he whom the Father raised from the dead ...constituting him judge of the living and the dead. Animated and drawn together in his Spirit, we press onward in our journey towards the consummation of history which fully corresponds to the plan of his love: 'to unite all things in him, things in heaven and things on earth (Ep 1:10).[15]

In this passage, we find a comprehensive cognitive meaning of the Incarnation, in contrast to its comparatively isolated treatment in the classic definitions of faith already referred to.

At this point, we can introduce the presence of other more tacit and pervasive dimensions of Christian meaning. They serve to put the objective affirmations of the Incarnation in a far richer frame of reference. These other dimensions of meaning, these other ways of meaning reality and responding to its meaningfulness, can be listed, first as *constitutive*, that is, as bearing on our living sense of identity; secondly, *communicative*, as establishing relationships of inclusion and belonging; and thirdly, *effective*, or world-transforming.[16] Let me illustrate each one in turn.

First, the constitutive function of meaning. To say it simply, we are constituted in a richer kind of human identity by our meaning-making. If I say that this dog is my pet, Pluto, I am not only meaning something about a dog as an object, but I am revealing, maintaining and confirming my own identity as someone who knows and loves this dog. I am constituted in a certain way. Similarly, if I say I am an Australian, I am not only saying that a country exists by that name, but that my national, cultural and historical identity has been constituted by such belonging, even if, in some foreign clime, I have only a passport to prove it. Further, the Australian constitution 'constitutes' me as a citizen with certain rights and duties within this country. By such obvious meanings, I own my identity in a certain way, and come to a feel, as it were, about myself and the world I live in, to present myself for what I am in a meaningful world.

Such constitutive meanings have been the special preoccupation of an existential form of philosophy, with the attention they have given to human consciousness and the self-constituting events and decisions that make us the kind of persons we are. They have highlighted the subjective pole of human experience. The question arises, then, as to what kind of subjectivity, what kind of lived identity does Christian Incarnational faith bring about? If I say the Word became flesh and mean it with the full assent of faith, I mean something; but I also have become someone in the process. I am 'In Christ' as St Paul would say. I have been constituted in the universe of meaning and of values by a distinctive feel for the universe and its mystery. As mystical theology would express it, the mystery of God revealed in Christ has become an indwelling reality in our hearts: the Christian is one who has been illumined by the light of the Word, as a member of the Body of Christ, as a temple of his Spirit, so to enjoy a liberated intimacy with the incomprehensible mystery, as the child of God.

In other words, the constitutive meaning of Christian faith is fundamental to its cognitive meaning, and often enough an impetus to its extension. To emphasize and promote this deeper meaning in terms of the transformation of human consciousness is, I would suggest, to enter more creatively into the contemporary world of meaning. It is to live with a sense of the universe as the unfolding of an all-enfolding and all-attracting love. Without this sense of self and universal belonging, we run the risk of being 'constituted'—of being made to see ourselves merely as the tiniest of cosmic fragments, or as one element in the biological chain, or as a stage in the evolutionary process that has no significance in a blind and aimless unfolding, or as a genetic mistake whose only saving grace is self-denigration for taking up such space in a world otherwise content to be without us.

Instead, in the constitutive meaning of Christian faith, human awareness unfolds into the peace the world cannot give, into the joy that no one can take away, as into the love that nothing in all creation can resist or subvert. Incarnational faith informs Christian consciousness with its meaning to make us participants in the genesis of the cosmos as the genesis of God in creation. It transforms the heart of the believer into intimacy with the whole mystery of what is coming to be. The field of such meaning is an atmosphere of hope in which the randomness, the chaos, the entropy of the physical universe is subsumed into the creativity of a larger grace, and the sins and failures of the past yield to the healing, forgiveness and transformation offered out of a limitless mercy.

Such a dimension of meaning marks the expression of the following passage:

> In reality it is only in the mystery of the Word made flesh that the mystery of the human truly becomes clear... Human nature by the very fact that it was assumed, not absorbed by him, has been raised in us also to a dignity beyond compare. For by his incarnation, he, the Son of God, has in a certain way, united

himself with every human being. He worked with human hands, he thought with a human mind. He acted with a human will and loved with a human heart. Born of the Virgin Mary, he has been truly one of us, like to us in all things except sin...

All this holds true not for Christians only but also for all people of good will in whose hearts grace is invisibly active. For since Christ died for all, and since in fact all are called to one and the same destiny, which is divine, we must hold that the Holy Spirit offers to all the possibility of being made partners, in a way known to God, in the paschal mystery...[17]

Here the accent is on the constitutive dimension of the meaning of the Word incarnate: 'life and death are made holy, and acquire a new meaning'[18]. Human consciousness becomes informed with a sense of dignity and destiny within what can seem a threatening cosmos: 'the riddle of suffering and death which apart from the Gospel overwhelms us'.[19] It establishes our identity in intimacy with the universal Origin: in the power of the divine Spirit we can invoke the ineffable mystery with a filial familiarity, as we share in the identity of the Son: 'we may cry out in the Spirit, "Abba, Father!"'[20]

But meaning is also communicative. Meaning functions not only by enabling us to mean *something* (cognitive), not only in making us a more meaningful *someone* (constitutive), but as a communication in a field of meaning. From such a field of communication, there emerges a community of language, thought, feeling, ultimate purpose and linked identity. At its most primitive and obvious, such a range of meaning is instanced as, say, when you can engage in conversation with someone. It is most noticeable in the formation of societies who 'socialize' their members into their respective meanings and values. On the specific levels of ecology and cosmology, we have a communicative meaning that sets us both in the larger community of life and in the vast process of the cosmos itself. In the ultimate horizon of our belonging together, Christian faith communicates a range of meanings that sets us in the community of creation, that identifies us as members of the body of Christ, that assures us of the shared 'Holy Breath' of the Spirit, that discloses our identity as sons and daughters of the 'Father'. Such an extension of meaning is well illustrated in the following words of Vatican II:

Just as God did not create human beings to live as individuals, but to come together in the formation of social unity, so he willed to make them holy and to save them not as individuals without any bond or link between them but to make them into a people who might acknowledge him and serve him in holiness... This communitarian character is perfected and fulfilled in the

> work of Jesus Christ, for the Word made flesh willed to share in human fellowship...
>
> ... As the firstborn of many brethren, by the gift of his Spirit, he established after his death and resurrection, a new brotherly communion among all who received him in faith and love; this is the communion of his own body, the Church, in which everyone, as members one of the other, would render mutual service in the measure of different gifts bestowed on each... This solidarity must be constantly increased until that day when it will be brought to fulfilment: on that day, humankind, saved by grace, will offer perfect glory to God as the family beloved of God and of Christ their brother.[21]

Here the accent is on the communicative meaning of the Word incarnate. It bears on the sense of solidarity with others, in their sufferings, joys and hopes. And today it is demanding an extension into the grandeur of a shared scientific story of our origins, and of our genetic interconnection with other life-forms of the planet.[22] When the earth itself has become the symbol of the one community of life, Christianity is invited to make deeper and richer connections with the ground on which it stands, with the nature in which it is immersed, with the cosmic body of its Lord.

Finally, it is a matter of recognizing more explicitly the effective dimensions of meaning. Such meanings transform the world in a certain way. The plans we make, the laws we enact and the ethical imperatives we follow, the technologies we design, the priorities we assign, the skills we employ, the cultural interests and political concerns we bring to any situation, all conspire into a world-making and world-transforming energy. Meaning makes the human world. The effective bearing of meaning could be illustrated in the recent emphasis on praxis in Political and Liberation Theologies. These modes of Christian thinking stress that, if the cognitive claims of Christian faith are to mean something in the real world, if the identity it promises is to be authentic, if the community it aspires to is going to be one of genuine solidarity, then its meaning must be effective. The test case of Christian meaning has typically been the genuine love of the neighbor. As this is extended into the whole neighborhood of creation in which such 'neighbors' live, the scope of the effective Christian meaning is expanded. The energies of love form the true face of the world. 'Love never ends' (1 Co 13:8). It is the ultimately decisive factor if the 'fruits of our nature and our enterprise' are to be purified, illuminated and transfigured:

> The Word of God, through whom all things were made, became

> human and dwelt amongst us: a perfect man, he entered world history, taking that history into himself and recapitulating it. He reveals to us that 'God is love' (1 Jn 4:8) and at the same time teaches that the fundamental law of human perfection, and consequently of the transformation of the world, is the new commandment of love...
>
> When we have spread on earth the fruits of our nature and our enterprise—human dignity, brotherly communion, and freedom—according to the command of the Lord and in his Spirit, we will find them once again, cleansed this time from the stain of sin, illuminated and transfigured, when Christ presents to his Father an eternal and universal kingdom...[23]

The more the Word is en-fleshed in the world of human meaning, the more all the dimensions of meaning come into play in a mutually conditioning manner. If meaning expands as an affirmation of the oneness of the universe in Christ, it illuminates our consciousness of being participants in a cosmic mystery of incarnation. If Christian meaning grounds a sense of inclusive belonging in creation, it inspires the transformation of the world itself. There is, if you like, a circulation—even an 'ecology'—of Christian meaning as faith seeks to understand its truths, its consciousness, its community and its vocation in each age.

5. The poem of the Word

To summarize, our reflections on the Logos incarnate in the cosmos have dwelt on four considerations. First, we considered how God's love for the universe is illumined by a recognition of the Word incarnate in the human world, and as profoundly affecting human consciousness: the Logos has entered the *dialogos* of the human conversation. Secondly, this led us to note the three arcs of the expression of faith, in its effort 'to word the Word' in the world of our experience. From there, we proceeded to the third consideration, how this focal meaning of the Incarnation as it was expressed in its classical doctrinal form now confronts Christian thought with new kinds of questioning. Finally, we indicated four dimensions of meaning in which this focal mystery is presently being expressed in Church doctrines.

Such an exploration of the dimensions of Christian meaning, and of the way it orders, suffuses, connects and orients our experience of the mystery of Christ, is well instanced in the visionary perception of Teilhard de Chardin as he writes:

> Throughout my life, by means of my life, the world has little by little caught fire in my sight until, aflame all around me, it has become almost completely luminous from within... Such has been my experience in contact with the earth—the diaphany of the divine at the heart of the universe on fire... Christ; his heart; a fire: capable of penetrating everywhere and gradually spreading everywhere.[24]

The brief points that I have made here not only sketch the dimensions of the great challenge confronting Christian thinking, but also provoke a re-reading of the past. The work of Irenaeus of Lyons and Origen are often instanced. But even in the work of such adversarial and logical thinkers as Tertullian, precious resources awaiting a larger retrieval are in evidence. To give one precious example:

> Think of God utterly taken up with the task of creation, with hand, sense, industry, forethought, wisdom, providence, and, above all, with that loving care which was determining the features. For the image of Christ, the man who was to be, was influencing every stage in the molding of the clay; because what was at that moment happening to the earth's clay, would happen again when the Word became flesh.[25]

Perhaps, when all is said and done, it is a matter of appreciating anew the Word Incarnate as a great poem. The sublimity of the hymnic prologue of John, referred to at the beginning of this chapter, points in that direction. The Word, the *Logos*, the Meaning Incarnate, works in human consciousness as a great poem. As Les Murray, a remarkable poet, has stated:

> Religions are poems. They concert
> our daylight and our dreaming mind, our
> emotions, instinct, breath and native gesture.[26]

Such words serve to remind us that the manifold meanings of faith are carried, not primarily in words or concepts, but in the pulse and momentum of our living. When the Incarnation is registered as the basic truth transforming our taste for reality and our whole feel for life as it carries us on, it will dawn in its inexhaustible meaning, for:

> Full religion is the large poem in loving repetition;
> like any poem, it must be inexhaustible and complete
> with turns where we ask, Now why did the poet do that?[27]

In the interests of 'full religion' or at least a fuller version of it, we have been involved in a kind of 'loving repetition' within the new contexts of our concern; and there are certainly those turns where we are left with a question for the divine Poet of the Word. While that Word draws us into a vision of all things held in existence, illumined and linked in the event of incarnation, there are moments of utter darkness, taking us to the depths of human tragedy, in the silence of the tomb in which the Incarnate Word is a tortured corpse. But we are not left there. No cosmic 'black hole' of meaninglessness has swallowed him. There is the 'white hole', if that is not too banal a parallel, of life transformed, inviting us into the realm of a new and final form of existence. There, the fragmented and groping meanings of faith come home; and what is obscure and unfinished on this side of darkness, blazes with light.

1. Note that the original Greek for what is here translated as 'world' is *cosmos.*
2. Material to debate such a position can be found in John Macquarrie, *Jesus Christ in Modern Thought*, SCM, London, 1990. Though there is no mention of anything like an 'ecological' attuned christology, there is some consideration of the 'Cosmic Christ', e.g., pp. 140–2; 167; 307f; 314–16; 428.
3. For a fuller theological consideration of the prologue, see John Macquarrie, *Jesus Christ in Modern Thought*, pp. 105–22. Macquarrie's own translation of this passage, especially his use of 'meaning' for *Logos,* is particularly imaginative. In view of what follows in my own reflection in this chapter, I find myself in considerable agreement.
4. For the importance of the symbol of light, see Jaroslav Pelikan, *The Light of the World: A Basic Image in Early Christian Thought*, Harper and Brothers, New York,1962.
5. St Augustine, *Confessions*, VII, 9.
6. I owe this notion, I think, to the great ecumenical theologian, Joseph Sittler. But, with the passage of the years, the precise reference now eludes me. For a more specifically Pauline treatment of these issues, see the concise but stimulating work, Brendan Byrne, SJ, *Inheriting the Earth: A Pauline Basis of a Spirituality for our time*, St Paul Publications, Homebush, NSW, 1990.
7. P. Teilhard de Chardin, *The Hymn of the Universe*, Collins, London, 1970, p. 168.
8. The particularly helpful example I find myself consulting is Denis Edwards, *Jesus and the Cosmos*, Paulist Press, Mahwah, New Jersey, 1991; also Jürgen Moltmann, *The Way of Christ: Christology in Messianic Dimensions*, SCM, London, 1990, pp. 274–312.
9. Cited in Christopher Mooney, *Teilhard de Chardin and the Mystery of Christ*, Collins, London, 1966, p. 136.
10. Let me note in passing, that, though I have been speaking in fairly classical terms of the incarnation of the Word, I do not mean to restrict this to the simple fact of the Word's becoming human. No Christian account of the meaning of the incarnation can pretend to any completeness if it leaves out the actual life, passion and death of Jesus, in his solidarity with the 'poor' and the hopeless. Nor can we speak of the Light that is overcoming the darkness without referring to the Resurrection, the final manifestation of the meaning of the Meaning, the *Logos*, uttered into hope's ongoing conversation. In later themes, we will refer to such aspects of the Christian mystery.
11. Here I touch on points explored at depth in the christology of Karl Rahner. Most relevant here is his seminal article, 'Current Problems in Christology', *Theologi-*

cal Investigations, I, Seabury, New York, 1974, pp. 149–200.

12. For an incisive analysis of this point, see Bernard Lonergan, 'The Absence of God in Modern Culture', in *A Second Collection: Papers by Bernard J. F. Lonergan, S.J.,* William Ryan, S.J. and Bernard Tyrell, S.J. (eds.), Darton, Longman and Todd, London, 1974, pp. 101–16.
13. One can easily overlook in this context the pioneering work of the Anglican theologian Lionel Thornton, *The Incarnate Lord*, Longman's, Green, 1928.
14. St. Thomas Aquinas, *Summa Theologica*, op.cit. 1, 43, 5 ad 2.
15. Austin Flannery, O.P., General Editor, *Vatican Council II: The Conciliar and Post Conciliar Documents*, Costello Publishing Co., Northport, New York, 'The Church in the Modern World', Par. 45, p. 947.
16. For these and many other aspects of 'the meaning of meaning', see the chapter entitled 'Meaning' in Bernard Lonergan, *Method in Theology*, op.cit., pp. 57–99.
17. Vatican II, 'The Church in the Modern World', op.cit., pp. 922–4.
18. Ibid., par. 22.
19. Ibid., par. 22.
20. Ibid., par. 22, end.
21. Ibid., par. 32, pp. 931f.
22. See Denis Edwards, 'The Integrity of Creation: Catholic Social Teaching in an Ecological Age', in *Pacifica* 5, 1992, pp. 188–94, for a fuller documentation of such communicative meaning.
23. Vatican II, 'The Church in the Modern World', op.cit., par. 38, pp. 937f.
24. P. Teilhard de Chardin, *The Divine Milieu*, Harper Torchbooks, New York, 1960, footnote 1, p. 46.
25. Quoted in Gabriel Daly, *Creation and Redemption*, Gill and Macmillan, Dublin, 1988, p. 77.
26. Les Murray, 'Poetry and Religion', in *Blocks and Tackles. Articles and Essays 1982–1990*, Angus and Robertson, Sydney, 1990, p. 172.
27. Ibid., p. 172.

SECTION IV

A Fourth Circle of Connections: From Within Creation

The mystery of creation is our second focus of connections.[1] As a noun, the word 'creation' signifies the totality of what God has created, is creating, will create. Moreover, there is nothing that is not God's creation, that does not owe its existence to the Creator. As a verb, it signifies the original and continuing activity of God's creating such a totality, whether the universe is considered in an evolutionary or in the comparatively static mode of classical philosophical theology. In both senses of the word—creation as a noun or as a verb—some notion of the Creator God is implied.

1. Knowing by negation

Before making a remark on the notion of God, it will be well to remind our imaginative propensities of their capacity to lead our minds astray very early in the consideration of creation. For instance, when we try to imagine creation as a noun, we naturally tend to imagine a totality of something to which we ourselves are oddly extrinsic: as though our thinking and feeling, praying, desiring and exploring were somehow a faint mode of being, built on the reality of creation, but not really part of it. Creation is thus imagined as a rather objective state of affairs to which we human beings respond. Here, I simply note that human response and responsibility are a dimension of the reality of creation.

Then, too, when we consider creation as a verb, God's original and continuous creating, creation can only be imagined as a very superior kind of doing something. The essential jolt to this imaginative fallacy was traditionally supplied by the addition of the mysterious little phrase *ex nihilo sui et subjecti*: the divine creative act presupposes nothing already in existence; no raw material; no chain of events; no previous dispositions.[2] God is not conditioned by anything already there, outside the Godself, to create. The imagination, however devout, cannot quite handle that. It tends inescapably to model God as the biggest factor in a world of doers and movers, in a cosmos of causes, principles and particular energies.

In contrast, the tradition of creation theology understands divine creating not as one category of causing or doing or deciding, but as a transcendent, therefore unimaginable, causality enabling creation not only to be, but to act. Creation in this sense means that God acts in the acting of everything, and causes in the causality of every agent. Creation strains the imagination to make the understanding recognize that we are dealing here with radical mystery. God is the ungraspable ground of all being and acting. These points we will further elaborate in due course. But first we must dwell for a moment on the notion of God.

It is tempting to discuss the perennial question of creation without using the time-(dis?)honored word, 'God'. The biblical and philosophical traditions that have formed Christian reflection into its sense of creation know other terms: the Word, Light, Spirit, Source, Life, Love, Be-ing (note, a verb, the Thomistic, *Ipsum Esse*—literally, 'Sheer To-Be'), Limitless Act (the *Actus Purus* of Western philosophic tradition), Ultimate Reality, Final Good, First Cause and many others. It is easy to feel that often even quite sophisticated discussion is blocked by the emotional fixations, fantastic distortions and conceptual straightjackets with which a variety of cultural conflicts have imbued this one word 'God'. Nonetheless, I am inclined to persevere with it. After all, it does not let us escape from the unredeemed elements in our history. The other terms still remain; and in dealing with the mystery that transcends human language, we need to keep in play the fullest kind of linguistic variations, even as we maintain the reverent restraint so characteristic of the Jewish tradition in which the personal name of God was seldom written or pronounced.[3]

The real problem, however, is not one of terms, be they marks on a page or vibrations of sound. It resides, rather, in how these terms are used, and in what they are supposed to mean. For the believer, the first and fundamental notion of God is found in an orientation to transcendent mystery which, in the deepest sense of the traditional term, alone 'saves our souls'. In more contemporary language, the sense of God is intimated as that which fulfills the movement toward self-transcendence: the meaning of God is, as it were, progressively anticipated in our search for ultimate and unconditioned meaning, truth, beauty, value, love and mercy; and in the divine self-disclosures with which life's questions have been answered.[4] Hence, however unformulated, the notion of God is tasted in the tang of life itself as it looks to its proper fullness. Unknown, yet ever attractive, such a mystery provokes surrender and adoration. God is the silence in which words fall away, the darkness where our brightest ideas fade in the presence of another light. God is experienced in the all-welcoming Love in which our restless hearts come home. In the divine realm of mystery, the final transformation hope tends to, is realized. It is the Space of ultimate connection and belonging. Hence to faith, God is first an invocable, all-including 'You', before the objectifications of 'It', 'He/She' come into play.

Not to recognize the anticipatory, notional character of the religious meaning of God is all too often to make dialogue with the reverently agnostic, be they scientists or meditative searchers; it is all too often a collision of congealed concepts. In such an unacknowledged struggle of fixed ideas, the scientifically enunciated version seems sophisticated, and the religious one, naive. This, I believe, is the problem which continually crops up in the matters we are about to discuss. So it is worth bearing in mind, in recognition of healthy biblical agnosticism, that God is not yet seen 'face to face', that God 'no one has ever seen', that God, though present to our love and faith and hope, is revealed only in a progressive darkness.[5] As affective and intellectual idols shatter, religious ideas are set free to point to the ultimate and original mystery rather than to grasp or contain it.

So with that little flourish of 'negative theology', let us go on to make some points on the religious doctrine of creation. Whether we speak of creation as a noun or a verb, such language operates in an horizon in which the One who is creator is present to the believing thinker: believers do not believe in the creator because they believe first in creation; rather they believe in creation because they believe in the One who alone can create.

Now, that is to admit that in the religious case, there is a certain undifferentiated affirmation of creator and creation before philosophical, historical or scientific questions are asked. As contexts develop, the creation question is framed to meet the demands of different mentalities—all the varieties of philosophy and psychology, of scientific method or aesthetic sensibilities, of evolutionary biology and quantum physics. If faith is to seek, or to find, further understanding, it has to wait on an extraordinary variety of highly differentiated contexts if it is to expand to its full intellectual and ethical potential.

2. Different starting points

Those who hold highly differentiated viewpoints of human intelligence without any consciousness of religious conversion ask the creation question, if they ask it at all, in a form quite different from those who have a specifically religious standpoint. If these latter tend to pose the question in terms of a prior 'faith seeking further understanding', or of 'faith making new connections', the former express their queries more in the terms of scientific understanding searching for faith of some kind; or at least for larger, ultimate connections. The implication is that the religious believer must wait on the findings of science or philosophy or history before belief can be possible, before God can be reasonably affirmed as compatible with some intellectual, moral or aesthetic scheme. The features of the resultant clash are clear: those who freely adore the ultimate mystery disclosed to their faith as the source and goal of all there is, are thought of as being merely 'conventionally religious'. It might also be suggested that the religious are necessarily threatened by other worlds of thought, since they cannot enter into them with the desired, critical integrity.

On the other hand, believers who might legitimately rejoice in an intimacy with the divine Ground of creation can often lack elements of an intellectual conversion. Through lack of leisure, or training or commitment, to say nothing of suspicion, fear or laziness, they can be unfamiliar with the real value of thought, philosophy and science. To them, the religious sense of creation is in danger of being replaced by some purely human method, from whose esoteric procedures they are barred. The universal expanse of faith, experienced in the democracy of an all-inclusive mystery of creation, is called on, they might feel, to yield a sophisticated elite presiding over the secrets of the universe. The believer is invited to become more critically distanced from the simple, living affirmation of faith, by becoming ever more believing in the assertions of science. Hence, the problem: at what point does credibility fade into credulity; and where does mass gullibility yield up the whole domain of human experience to the mathematical physicist, for example, who may have some tentative place for the mind of God disclosed to his research and hypothesis, but who can speak of the universe without any mention of death or love, of art or grace, of scholarship or revelation, of beauty or morality. Such is the problem conveniently instanced in Paul Davies' latest book, *The Mind of God*.[6]

The question of creation is, then, an area where deep methodological questions about the collaborative character of human knowledge are posed. Is there a way of respecting the whole range of the human conversation about meaning, truth, value? About the universe of all that is? About the mystery of its origin? Is there any point in talking about creation, and of the divine origin of all things (including the human mind and heart), if we take no care to be systematically open to all the data, above all the data which each of us is, in the complexity of our conscious living? For we are each alive, conscious, plunged ecstatically in a world of daily meaning, truthfulness, responsibility. Here love illuminates the best of our lives, just as beauty in art or nature continually refreshes our perceptions of the uncanny occurrence of our existence.

If what I am saying sounds too broad or obscurantist to the hard-nosed scientist, I have not the slightest intention of denying the value of scientific research. By implying that the totality of reality eludes any exclusively scientific method, I am attempting to locate scientific method in a larger field of human experience and exploration. Indeed, I think that the theology of creation points to a background theme in which all our diverse human creativities can improvise their variations in the expression of the really Real. We can anticipate the direction of our reflections by citing the following words of Michael Polanyi:

> Admittedly, religious conversion commits our whole person and changes our whole being in a way that an expansion of our natural knowledge does not do. But once the dynamics of knowing are recognized as the dominant principle of knowledge, the difference appears only as one of degree... It establishes a continuous ascent from our less personal knowing of inanimate matter to our convivial knowing of living beings and beyond this to knowing our responsible fellow men. Such I believe is the true transition from the sciences to the humanities, and also from our knowing the laws of nature to our knowing the person of God.[7]

3. From the biblical to the modern question

The unique transcendent act of God in calling things into existence is expressed in the Hebrew term *bárá.* It seems clear from the writings of Deutero-Isaiah and Ezekiel that the great theme of God as creator is employed to bolster the hopes of Israel in its experience of captivity. God can bring about a new exodus; God can bring about the radically new because he is the creator, because he is dependent on nothing; because he is sheer originality in regard to all reality and history.

The two creation accounts which we meet in the early chapters of Genesis are not expressed under the same degree of historical pressure.[8] They represent a marvelous morning dream into the universe as God's original gift. The world appears as an ordered but differentiated whole. Humanity is given into such a totality to share in the divine creativity through responsibility and care. Within such a world, the human community will find nothing to equal God, for God is the world-originating source of all good. Though evil has its opaque presence in human experience, it is not from God; evil is from creatures gone wrong; it is the perversion of creation, neither its original nor final condition. In every reality and behind every moment of history is the utterly free and unconditioned power of divine creativity.

While these and related themes are familiar,[9] the sense of creation that Christianity inherits from Israel would be truncated if no mention were made of the more immanent, participative emphasis evident in the Wisdom literature. What we find there contrasts with the rather hierarchical and pictorial accounts characteristic of Genesis. If these accounts are expressed in terms of visual and auditory imagery, the sensory texture of the Sapiential experience of creation is more kinetic and tactile. Most obviously, the feminine character of creative wisdom is stressed, as the divine presence pervades and encompasses all experience. To give one example:

> ...for wisdom the fashioner of all things taught me. There is in her a spirit that is intelligent, holy, unique, manifold, subtle, mobile, clear, unpolluted, distinct, invulnerable, loving the good, keen, irresistible, humane, steadfast, sure, free from anxiety, all-powerful, overseeing all and penetrating through all spirits that are intelligent, pure and altogether subtle ... she is the breath of the power of God and a pure emanation of the glory of the Almighty...(Ws 7:22–25).

By evoking the translucent freshness of reality as divine creation, such a passage suggests the kind of transformation that occurs in the human mind and heart when God is acknowledged as creator, and when the world is recognized as God's creation. The divine presence is to be found in and through the whole order of creation as an all-attractive, pervasive, life-giving reality (cf. Pr 8: 22–9:6; Si 1:1–20; 24:1–22; Ws 7:22–9:18; cf. Ep 4:6). Thus we have an indication of Israel's true *philosophia*, the 'love of wisdom', as it relishes all reality in its God-given freshness, in each present moment. We must wonder about the extent to which the neglect of such themes related to biblical Wisdom has left theology somewhat impoverished and awkward when it comes to face the great ecological and cosmic questions of the day.

We are taken closer to the nub of the contemporary question of creation in the following marvelous passage. Here, the biblical sage reflects on less conclusive kinds of wisdom:

> ... they were unable from the good things that are seen to know the one who exists, nor did they recognize the artisan while paying heed to his works; but they supposed that either fire or wind

> or swift air or the circle of the stars, or turbulent water or the luminaries of heaven were gods that rule the world. If through delight in the beauty of these things people assumed them to be gods, let them know how much better than these things is their Lord, for the author of beauty created them. And if people were amazed at their power and working, let them perceive from them how much more powerful is the one who formed them. For from the greatness and beauty of created things comes a corresponding perception of their creator. Yet these people are little to be blamed, for perhaps they go astray while seeking God and desiring to find him. For while they live among his works, they keep searching, and they trust in what they see, because the things that are seen are beautiful. Yet again, not even they are to be excused; for if they had the power to know so much that they could investigate the world, how did they fail to find sooner the Lord of these things? (Ws 13:1–9).

Today 'the good things that are seen' have been immeasurably extended. The magnitude of the energies and the infinitesimal intricacies of pattern and design are so fascinating, and so subversive of former visions of reality that scientific language verges on the religious in its effort to express the numinous value of the cosmos it celebrates. On a purely scientific level, a new enchantment with the beauty and wonder of the cosmos is apparent. Little wonder that there are those who feel they have discovered 'the gods that rule the world'. In comparison, the God of traditional faith seems little more than a naive projection.

We find ourselves in an exuberant Babel of languages: different minds and different methods seek to name the ultimate, and a new holistic concern tries to overcome the splintering and reduction of the former perceptions of science and religion. Still, the biblical sage graciously admits that those who confuse the wonders of creation with the wonder of the Creator are 'little to be blamed'. A realistic search is in evidence: if the real wonders of the world 'get one in' and 'turn one on' to a confidence in the beauty, complexity and sheer uncanniness of it all, it is likely that the search will continue for the full mystery intimated in such experience.

And yet there remains the possibility of stopping short: '... how did they fail to find sooner the Lord of all things?' Further questioning can be silenced with all the versions, both ancient and modern, of Carl Sagan's proclamation: 'The cosmos—as known by science—is all there is, all there was, and all there will be.'[10] Or is the bleaker statement of the geneticist Jacques Monod the untransgressable point: 'The ancient covenant is in pieces: man at last knows that he is alone in the unfeeling immensity of the universe out of which he emerged only by chance.'[11] For others, the tentativeness of further searching is evident, as when Werner Heisenberg wrote, 'Although I am convinced that scientific truth is unassailable in its own field, I never found it possible to dismiss the content of religious thinking ... Thus, in the course of my life, I have been repeatedly compelled to ponder on the relationship of these two regions of thought.'[12]

Perhaps it would not appear as strange to the biblical sage as it does to a modern positivist mentality, to witness the recent tendency of scientists to rehabilitate the question of God in the realm of science itself as it tries to explain the uncanny design of the universe. Paul Davies, for example, is a leading popularizer of this shift:[13] 'It may seem bizarre, but, in my opinion, science offers a surer path to God than religion'.[14] The 'bizarre' character of such an assertion might be more vivid to the modern scientist than to the most venerable philosophical tradition of the West. As Etienne Gilson remarks, himself quoting the scientist Louis de Broglie:

> ... 'The great marvel in the progress of science is that it reveals to us a certain concordance between our thought and things, a certain possibility of grasping, by means of the resources of our intelligence and the rules of our reason, the profound relations existing among phenomena ...One cannot but be astounded enough at the fact that some science is possible'. If one thinks that philosophers have been amazed at this since the time of Plato at least, that is to say, more than twenty centuries, these words take on unexpected meaning and relief. They mark the moment when science itself recognizes the mystery which lies over its existence. The fact is even more worthy of mention in that L. de Broglie's remark joins that of Einstein, ' ... the most unintelligible thing on the subject of the world is that it is intelligible.' [15]

The 'bizarre' nature of Davies' remark seems quite benign, in fact, when you take in the whole philosophical tradition stemming from Plato (in the *Timaeus*), and Aristotle (as in his *Physics* and *Metaphysics*). And though medieval theological tradition would admit a whole range of complex distinctions between knowing God by reason and assenting to the divine mysteries in faith, its peculiar brand of hardy intellectualism would not be altogether amazed by what Professor Davies has to say. The variety of medieval efforts to 'prove the existence of God' from the fact, the structure, the movement of the world of our experience, together with their reflections on created reality as exhibiting a 'vestige' of the divine (as in the work of Aquinas and Bonaventure), suggest as much.[16] More obviously, and this is more surprising to the post-Newtonian science of today, the scientific approach to God accords with the conviction of Newton himself in the 28th query at the end of the *Opticks*: 'The main Business of natural Philosophy is to argue from Phaenomena without feigning Hypotheses, and to deduce Causes from Effects, till we come to the very first Cause, which certainly is not Mechanical'.[17]

Admittedly, as we mentioned above, there is a certain limitation in searching into 'The Mind of God' when no mention is made of love or death, of value or ethics, of art or grace as divine self-revelation; or for that matter, of the problem of evil itself. The mind of God cannot be explored with complete profit if the mind of the explorer is so mathematically inured against the drama and struggle for real life. Such an exploration can look like a sophisticated escape mechanism evading the intellectual demands of 'the whole issue' of existence. The result is that anyone more dramatically exposed to the tragedy and wonder of life is tempted to dismiss exclusively mathematical explorations of God as the productions of autistic savants.

But it need not be so. The very fact that so much of the cosmos is found to be so profoundly intelligible and so elegantly beautiful, luminous with that *splendor entis*, that 'radiance of being' of which Aquinas speaks, invites the mind into its ultimate adventure: an exploration of the original mystery of it all: 'if they had the power to know so much... how did they fail to find sooner the Lord of all things?' The gently insistent question posed by the biblical sage is met with the declaration of Judith Wright, one of the great poets of our time:

> My search is further.
> There's still to name and know
> beyond the flowers I gather,
> the one that does not wither—
> the truth from which they grow.[18]

This, in turn, links back to a beautiful expression of medieval wisdom:

> Creatures as far as in them lies do not turn us away from God but lead to God. If they do turn us away... that is due to those who, through their own fault, use them in ways that are contrary to reason.[19]

4. Image, thought and exploration

Imagination, however necessary for understanding, is not understanding. While images provoke and focus thought, they do not replace thinking. Hence, the great theological and philosophical doctrines of creation in an obvious sense go beyond the biblical and creedal imagery of God as maker, actor, artisan, potter, and so forth. Philosophy and theology arise when the undifferentiated imaginative life of faith makes contact with the world of theory, be it in philosophy or science.

To explain further. The biblical narratives and confessions, their summation in the creed—'We believe in God, the Father almighty, creator of heaven and earth, and of all things, visible and invisible'—appeal to the drama and vitality of religious consciousness. To the believer, everything is God's creation. The creator, the origin and goal of everything, acts in and through everything and all the events of history. Because God is the origin of all, there is a fundamental unity of all things in that one creation. There is nothing outside the creative scope of God. Within this oneness of creation, humankind has its special place as the image of the creator God, as the worker in the garden of creation, as God's steward caring for all that has been declared good, or as the builder of the saving ark. For human consciousness to be fixated in anything that God has made is to project an ultimacy onto the creature that it cannot bear: this would be idolatry, the radical displacement of mind and heart into anything less than total meaning and ultimate value. To the biblical mind, idolatry is the most characteristic perversion of the human spirit.

Now, the philosophical current of creation theology, though it does not disown either vivid religious conviction or the metaphorical expressions characteristic of religious consciousness, operates on a more abstractly metaphysical level. Here, metaphors are understood to be metaphors;[20] and analogies, however necessary they are to affirm transcendent mysteries, must include a moment of negation: God is 'good'—though 'not good' as limited entities are good, but in the 'eminent' manner of the divine mystery. The negating moment occurs at the heart of any such analogical knowledge.[21] It makes for a space of reserve and silence in the theological affirmations. In the present instance, we need to bear in mind that no theology can express precisely either what the creator is, since the divine is beyond any worldly meaning; nor what creation is as a verb, since the divine creative act is not like any familiar instance of 'making' (which presupposes something, however potential, already existing); nor even creation as a noun, since the utterly fundamental character of the total dependence of the universe on God for its existence is beyond any other relationship we can know. Those three notions—the creator, creation as a verb, creation as a noun—however much they arise out of the concreteness of existence, lead to a point where no language nor any thought is really adequate.

Such a reserve must carry over into any dialogue with contemporary science. True, we find ourselves in a new historical context affected by new scientific methods, and more profoundly, by new mentalities or differentiations of consciousness, as scientists locate themselves in new ways in the universe of their explorations, freshly alert to the relativity of their standpoints. Nonetheless, whatever its range of reference, and however much the subjectivity of the scientist is inextricably bound up with the objectivity of his or her judgments, scientific research is like any other kind of human knowing. It deals with some specialized band of data, forms it into imaginative models, locates its questions in a world of meaning, weighs the evidence for the most probable position, and seeks applications of its findings in accord with a certain scale of values.

Such procedures also structure the explorations of theology. The theological difference lies, of course, in the kind of data: those pertaining to religious experience as it unfolds through history. Theologians formulate a doctrine of creation out of a radical, continuing demand for ultimate meaning of such experience. They can never pretend to the possession of some final answer, as though the mystery of God were comprehended. It is more a matter of coming to limits, of experiencing the continuing force of the ultimate question, even while any ultimate answer remains locked in silence and darkness. The aim of theology, then, at least in its philosophical form, is to turn the inquiring mind from the reasons sufficient for any judgment of truth to the sufficient reason for sufficient reasons; from the meanings we find in a meaningful world to the meaning of all its meanings.

The doctrine of creation looks more and more like a question. It unfolds as an open structure, open to the source of existence without comprehending it. Hence, the nature of God remains radically unknown—or known only in the intelligibility, the meaning of the universe. So, to affirm that the universe is God's creation is to lead everything back to its ultimate mystery: *reductio in mysterium*.

In its overture to original mystery, the theology of creation resists all forms of reductionism to something less than ultimate truth—any tight little control that works like a defence mechanism protecting the prestige of one area of expertise, be it physics or chemistry, biology or sociology, psychology—or in theology itself. The reductionist declares the 'really real' as 'nothing but' a flow of electrons, or chemical reactions, or biological drives, or cultural projections, or psychological needs, or social symbols; or, as in the case of the religious version of reductionism, nothing but the action of God. The reductionist mentality blooms into the many forms of fundamentalism, the 'nothing but-tery' of minds refusing the whole promise of their intelligence.[22]

The doctrine of creation, both in its classic metaphysical form and in its contemporary familiarity with the scientific manifold of our experience of the world, resists such reductionistic fundamentalism. It insists that the mind stay open and that the heart not rest until the total explicative mystery of it all is revealed. Before moving to the contemporary context of the question, let us sample something of the classic medieval account of the meaning of creation. In many ways, it is a surprising resource.[23]

5. Pointers from the past

In *Summa Contra Gentiles*,[24] Aquinas asks whether the consideration of creatures is useful for the enlightenment of faith. His positive answer reveals a sturdy and reverent realism: first, he suggests, by considering the varied elements of creation, all of which image forth God in some way, we come to a more intense and concentrated sense of the divine wisdom. It is as though the created universe of particular beings draws the exploring mind into a more inclusive horizon, in the expanse of which the creative wisdom of God can be more realistically appreciated. Similarly, it engenders a sense of varied creativity of the divine power at work. But his third reason is more fully expressed:

> Thirdly, this consideration of creation sparks in the human soul a love for the divine goodness. Whatever goodness and perfection is spread out in different creatures, it is a totality brought wholly together in him as in the source of all goodness... If therefore the goodness, the beauty and the delicacy [lit. *suavitas*, sweetness] of creatures so lures the human soul, the source-good of God himself, diligently compared to the streams of goodness in the variety of creatures, inflames and attracts totally to itself our human souls.

Here the divine is experienced in creation as an attractive force progressively luring human consciousness into the experience of the *fontana bonitas*, the fontal goodness that God is.

His fourth reason appeals to a more intimate sense of participation in the divine. By considering the variety of creation, we human beings achieve in ourselves 'a certain likeness to the divine perfection and wisdom'. Since God knows (and loves) everything in himself, and as we human beings participate in God's self-knowledge and love through faith, we begin to share in the divine consciousness of creation. To illustrate this luminous new consciousness of creation, Aquinas goes on to quote St Paul, 'And all of us, with unveiled faces, seeing the glory of the Lord... are being transformed into the same image ...' (2 Co 3:18).

Given such an appreciation of the created world as a manifestation of the creator and as a participation in the creative mystery, one must cherish in such a vision a classic resource for the formulation of contemporary ecological and cosmically attuned theology. Nonetheless, as a corrective to any form of naive holism, there can be no question of confusing creation or nature with God. In fact, Thomas goes on to say that the proper appreciation of creation is necessary not only for finding truth, but for excluding errors[25]: by intelligently searching into the world of creation, we find that no creature is independent in its existence, and that no element of our world is 'the first cause'. Hence, it becomes a matter of valuing the variety of created realities within the englobing mystery of existence as a gift. In this medieval vision, the variety of creation radiates from the one inexhaustible divine origin:

> God is... the most perfect agent. Therefore, it belongs to him to induce his image in created things in the most perfect way in a manner that befits created nature. But created things cannot attain to the perfect image of God in a single form: the cause exceeds the effect; for what exists in the cause in utter simplicity, is realized in the effect in a composite and pluriform manner... It is fitting, then, that there be a multiplicity and variety in created things so that God's image be found in them perfectly in accord with their mode of being.[26]

Such a passage indicates a theology of the variety and precious particularities of creation. Each element of creation has its own quasi-absolute value. It is a particular realization of the divine. Perhaps the numinous sense of nature that so many ecologists bring to their commitments today is living from a subterranean connection with this great medieval vision, for:

> God planned to create many distinct things in order to share with them and reproduce in them his goodness. Because no one creature could do this, he produced many diverse creatures, so that what was lacking in one expression of his goodness could be made up by another; for the goodness which is simply and wholly in God, is shared in by creatures in many different ways. Hence, the whole universe shares and expresses that goodness better than any individual creature.[27]

And yet there is a further point, too often overlooked, in this classic sense of creation. The mystery of creation, in its variety and unity, not only shares in the beauty and goodness of the creator. It is brought into existence by an act of free creative love: Aquinas builds up to this conclusion in the following passage:

> God loves everything that exists ... The will of God is the cause of everything; and so it follows that insofar as anything exists or is good under any aspect, it is willed by God. To whatever exists, God wills some good. So, since love means nothing but wishing

> good to something or someone, it is clear that God loves everything that exists. But not in the same way as we human beings do. Because our will is not the cause of the goodness of things, but rather is moved by such good as an object of desire, so our love, by which we wish good for another, is not the cause of the good of that person. But rather, such goodness, real or imagined, provokes the love by which we will the good to be kept that is already possessed, or the good to be added that is not had. We work to that end. *But the love of God is infusing and creating the good in everything.*[my emphasis][28]

The being and variety of creation is total gift. God's love is not first attracted toward something already existing. Rather, everything inasmuch as it exists, is loved into existence by the divine freedom: *amor Dei infundens et creans bonitatem in rebus.* Creation, from this point of view, is enacted as a communication of the divine joy in the existence of what is other. It is sheer gift: *bonum est diffusivum sui* (the Good is diffusive of itself). Hence the universe is a divinely chosen order of being, and not a kind of necessary emanation from, or completion of, a remote deity. In all its variety and connectedness it is a communication from the heart of God.

The stable world of particular natures linked in a great chain of being is, indeed, different from the evolutionary universe of today. In contemporary thinking, the whole is a process before it is an ordered collection of particular natures. But here, I think, two remarks might be made.

First, Aquinas' serene delight in the specific value of each element of creation as loved into being, and serving to manifest the divine goodness in a particular way, mitigates the cosmic sadness of an uncritical evolutionary myth. In its sense of the past, such a myth reads like an obituary of the casualties of evolution. The fittest survived; the weakest did not; and even what survives is subjected to the blind, harsh law that values only the future. In contrast, the medieval vision, even while it leaves unknown the mysterious ways of Providence, finds an existential value in things simply because they existed—or existed in a certain way at a certain time. It is a reminder to those who would hurry to frame the laws of evolution to allow space in their theories for two considerations.

First, each individual entity is a world of mystery in the sheer fact of its existence. Before it can be considered a link in the chain of evolution, or as an aspect of the larger emerging complexity, it is, or was, *there*! The evolutionary potential—or lack of it—of the *Marella splendens*, or of any of eighty thousand extraordinarily complex creatures unearthed in the Burgess Shale in Western Canada, does not evacuate the fundamental wonder of its existence in the play and contingency of what happened in time.[29] The unique existence of the individual entity is so often 'the missing link' in evolutionary thinking.

Secondly, the seemingly purposeless variety of what once existed or is existing in its unique manner should make us wary of any premature closure on the whole story of what is going on. The human mind is not a detached spectator, but is part of the emerging process. Our reconstructions and extrapolations, whatever the progress, access only fragments of the meaning of the whole process. The Thomistic emphasis on the necessary plurality of creation and on the value of each existent not only contests the simplicity of evolutionary myths, but serves to keep evolutionary theory attentive to a truly inclusive wholeness. The system must remain open. The full story awaits, in patience and tentative exploration, the full understanding of creation that resides only in God.

My second remark accords more with the evolutionary worldview, and hence implies historical limitations in the 'good natured' universe of Aquinas. For the good that God is creating is not yet fully realized. The whole 'groaning' (Rm 8:18–28) totality of creation is awaiting its liberation. Our human role places us in creation as free agents, collaborating with the divine creative energy at work. Human freedom is not in competition with God, but God's most promising creation within the created world. Human history is destined to work with God to bring about a good that is still emerging. As human existence participates in the divine knowledge and love, it turns 'an incomplete and seemingly futile universe toward construction of a world that is good'.[30] In this more eschatological perspective, human freedom not only bears the guilty burden of appalling destructiveness; it is the condition of the good, the happy outcome, the efflorescence of love in the final ecology of the world.

But that is to anticipate. The reflective faith of Aquinas moves in a universe redolent of the goodness, beauty and simplicity of its Origin. To identify the universe as God's creation is to appreciate it 'in depth'. Yet the metaphysical language and the Ptolemaic models of its expression are oddly abstract compared with our modern apprehensions of reality. So we ask, what are the characteristics of this new worldview?

6. Dimensions of creation today

A sense of creation today has to include new data in its vision, data hitherto not only inaccessible but unimaginable in the past. The psalmist of old saw, 'The heavens declare the glory of God, and the firmament shows forth his handiwork' (Ps 19:1). Solomon's wisdom was limited to fire, wind, swift air, the circle of the stars, the luminaries of the firmament. But the might of thunderstorms, the light of the sun and moon, the twinkling stars, the steady glow of the planets, the shimmering Milky Way are only a tiny fraction of the realities that intrigue the modern mind. For instance, today the glory of the heavens includes the icy rings encircling Saturn, Neptune's mysterious satellites, the delicate blue of the earth seen from the moon, our one hundred billion sun galaxy in a universe of at least one hundred billion galaxies. Our telescopes and computations extend to the ultra-energetic quasars, to explosive enormous supernovas, even to the space-time rifts of 'black holes'. We know, too, that a million earths could be hidden in the sun; that light blazing through space at 300,000 km per second takes 100,000 years just to span our particular galaxy. Though Alpha Centuri is our nearest star, it is still four light-years away. From the quark, with its eighteen variations, to the quasar, it is a universe of incredible proportions.

Today's sense of the immeasurably great and the immeasurably small dimensions of physical matter has to include, too, hitherto unimaginable expanses of time; and within that time, the evolutionary dynamics that have produced this human world. Human consciousness, trembling before the uncanny extent of the fifteen billion year history of this universe, comes to realize that these thousands of millions of years are, quite literally, *our* past. All its seemingly chancy gropings have come up with what now is. In some uncanny sense, the universe, from its beginnings, 'knew we were coming'.

As one philosopher of science puts it, 'if the universe were in fact different in any significant way from the way it is, we wouldn't be here to wonder why it is the way it is'.[31] In the first few seconds of this 15 billion-year-old universe, there appeared all the fundamental particles and constants without which life on this seemingly insignificant little planet would have been impossible.[32] The velocity of light, the charge on the electron, the mass of the proton, the constant emissions of energy, are exactly what they should have been if life were to teem in all its forms on earth. A slight difference then would have meant no life now. If the charge on the electron had been slightly different, if the reaction between two protons had been slightly different, if the force of gravity had been slightly different, then the universe as we know it would never have come to be. All those millions of hydrogen atoms formed fifteen billions years ago, which today in various combinations make up the composition of the human body, would have turned into the inert gas helium. There would have been no water, no life, no you, no me... [33]

The chancy, incredibly 'iffy', uncanny present of our world depends on an expansion rate of infinitesimally calibrated precision. In his *A Brief History of Time*,[34] Stephen Hawking estimates that if the rate of the expansion of the universe one second after the Big Bang had been smaller even one part in a hundred thousand million, the whole would have collapsed before it reached its present size.[35] But if the rate of expansion had been greater, the universe would have become an enormous bubble of gas blown out too quickly to allow the stars to form. And if there were no stars, there could be no formation of the heavy elements that are essential for the biological processes characteristic of this planet.

In short, if the universe had been too much in a hurry to be merely big, it would have had no time for life to happen. Therein lies, perhaps, a parable basic to all existence. Even if 'small is beautiful' might not be a universal axiom, beauty does cosmically consist in the right proportion.

The cosmos now appears as an immense field of energy, structured in terms of four forces.[36] A strong force holds protons and neutrons together in atomic nuclei. If there had been a one per cent difference in its strength, there would have been no carbon formed inside the stars, and consequently no material for the production of DNA molecules. A weak force allows for radioactive decay. If this force had been even slightly stronger, all hydrogen would have dissolved into helium: with the result that there would be no water, and even no sun to warm things into life on earth. An electromagnetic force orders the path of light and the behavior of charged particles. If this had been stronger, the stars would have been too cold to explode as supernovas, and so to cast forth their heavy elements to be the raw materials for life's emergence. Then, too, the force of gravity holds all physical entities together in fields of mutual attraction. If that had been different, the other forces would have acted in different ways, and the present happy outcome would have been precluded by a heaven of cold stars, or an inferno of billowing gases.

Again a cosmic parable. If the universe were a world of unrestrained force, it would have had no place for life.

Further, it is estimated that, in the constitution of matter, there is a tiny asymetrical ratio of particle and anti-particle. If, for example, every proton had been perfectly matched by an anti-proton, there would have resulted a kind of mutual cancellation. But there is the faintest edge of excess breaking the symmetry, even if this is computed in terms of one in a billion. It is this minute excess, this infinitesimal asymetrical surplus, that becomes a window of opportunity for the emergence of the universe as we now know it.

Again the cosmic parable: if the universe had been intent on mere balance, closed against all conflict, repressive of any excess, nothing could have happened.[37]

And yet, here we are, in this moment of time, alive on this planet, immersed in this universe, just as both earth and universe are alive to themselves in us.[38] Our bodies, minds and hearts are part of an awakening cosmic mystery. Paul described the whole of creation as groaning in one great act of giving birth. He saw us human beings as groaning too, for the fulfillment that is not yet; more mysteriously, he understood the creative Spirit of God groaning within us to inspire hopes worthy of the mystery at work (Rm 8:18–28). Such a vision is not far removed from that of Eric Chaisson of Harvard:

> We are not independent entities, alien to the earth. The earth in its turn is not adrift in a vacuum unrelated to the cosmos. The cosmos itself is no longer cold and hostile—because it is our universe. It brought us forth and maintains our being. We are, in a literal sense, 'children of the universe'.[39]

In other words, the sense of creation offered to us today brings together an appreciation of how we human beings are from God—'children of God' in the biblical sense—and 'children of the universe', born of the earth. We are made in the image of God, and yet we are earthlings: we live only in a genetic solidarity with myriad other forms of life on this planet. If they serve and sustain us, we are called to a responsible stewardship of them. We human beings are related in a web of life with some millions of other species. Waratahs and wallabies, kookaburras and king prawns, frill-necked lizards and ferns—in the one web of life, they are all our relatives in a wondrous cosmic solidarity. Elements of the stars are in the phosphorous of our bones. The same hydrogen which makes the stars burn, energizes our bodies and powers our imagination.[40]

So much has changed as knowledge has expanded. Still, it is worth noting that in the biblical vision of creation, other and higher forms of life were presumed. Their presence is now largely mediated to us only in the various liturgical prefaces in which the assembly is invited to join with 'angels and archangels, cherubim and seraphim' in a cosmic hymn of praise... At least to evoke that forgotten dimension of the biblical cosmos suggests a more wonderful view of the living universe than that of the flat materialist anthropocentrism that has got us into so much trouble. The cosmic feelings and symbols of solidarity with all that is both below and above our form of life cannot but promote a greater awe and fresh questioning in the presence of so much that is still unknown. Such largely ignored elements of the tradition still have their power. They function as symbols, to say the very least, of dimensions of reality that have yet to be considered in our efforts to understand the commonwealth of life and the community of consciousness in the universe of our present perceptions.[41]

In short, our present perceptions of time and space, of the particular evolutionary dynamics that science can discern, of the whole emergent process of universe preceding any structure of determined natures all provoke a fresh posing of the age-old question of creation. From one authoritative source comes the question:

> If the cosmologies of the ancient Near Eastern world could be purified and assimilated into the first chapters of Genesis, might contemporary cosmology have something to offer to our reflections upon creation? [42]

7. A refinement of method

Even if our answers to such a question are positive, a newly expressed theology of creation has to be acutely aware of the profound cultural shift that has occurred. To stress yet again, the doctrine of creation has been in no small way dependent on a classical form of culture dominant in Western history until recent times. For example, some notion of God was culturally available: atheism was simply not an intellectual option. Then, there was the common inheritance of philosophical categories—causality, matter and spirit, the variety of relationships, the distinction between living and non-living things in an ordered, hierarchically differentiated universe. While the classical philosophical tradition went 'beyond physics' (*meta-physics*) with its sense of the mystery of being, and its elaboration of the attributes of God in a negative theology, it lived to a large degree in a picturable universe. It envisaged a hierarchy of forms in an ordered whole. Reality was predictable because of the natures of things. There was a natural law resulting from the participation of everything in a structured organic whole—though and in the last few centuries, at least, with Newton and Descartes, natural law tended to become more and more unnatural as the model of the universe as a machine became dominant.

Only with difficulty (here the complex history of the condemnation of Galileo is the dramatic instance) did philosophical theology get beyond a geocentric to a heliocentric universe. Its Aristotelian ideal, however remote, was certitude, the explanation of all things according to their causes. Such was deemed a possibility because of the ordered metaphysical world that was presumed. Everything was analogically related in the one universe of being; everything participated in the one all-creating mystery—God, sheer Be-ing, Pure Act, the First Cause.

In such a universal order, human beings were actors in the ordered universal of existence. The 'why' of it all was the pre-eminent issue. The deep context was the religious and spiritual experience of human beings, valued and available dimensions of human existence.

Contemplative reflection was the path to truth. Today, that quiet, reflective path has yielded to the rigors of scientific method.

In contrast to such an ordered, participative, God-deriving wholeness, modern scientific culture is often diagnosed as suffering from the 'absence of God'. Not dependence on the divine, but the autonomy of freedom and intelligence are the prized values of this post-Enlightenment world view. Modern scientific culture characteristically appeals to empirical data and verifiable experiment. Its aim, in all the variety of its methods, is to give an account of the accumulated sense-experience of the world. Here, the human person was less an actor participating in universal ordered existence, and more a spectator analyzing and controlling reality from a particular empirical access to a limited band of data. Scientific knowledge confronted the world rather than participated in its meaning.

The result, predictably, was the reduction of the metaphysical to the physical, and to levels of the physical in which any particular science was expert: reality is 'nothing but...' this or that material, physical, chemical, genetic reality.[43] Modern science focused on not 'why', but 'how' things came to be. In such a context, God, at first, then at best, is identified as filling the gaps when no data is available, or as being the explanation when no suitable experiment can control the reality in question. Thus, in a scientifically reductionist method, 'God' is by degrees edged out of human experience, and the only truths accessible to the human mind are those declared as verifiable in modern experimental techniques: God is neither found by a surgeon's scalpel, nor seen through a scientist's microscope, nor required as a factor in the mathematician's formula, nor identified as a component in the world system.

What is more, to the degree the notion of causality figures in modern scientific method, it is radically different from its metaphysical antecedents. Up to the beginning of this century, as we mentioned above, the world of nature was regarded as determined and fundamentally, even mechanically, predictable. Today indeterminacy is recognized at the microlevel; and, because of the complexity of the causal chains and links, unpredictable at the macrolevel. This causal uncertainty is compounded in the megaworld of intergalactic distances and cosmological processes unfolding over billions of years. While a causal, certain account of reality might remain as a remote ideal, scientific methodology tends pragmatically to limit itself to techniques of correlation of phenomena as it surveys the limitless number of situations and events making up the physical world. Thus, it largely prescinds from the older causal language. Its aim is more one of approximation, the convergence of probabilities, the testing of hypotheses.

The making of wholesale causal connections will tend always to seem premature, and notions of metaphysical causality the sediment of the naive realism of a prescientific era. The past picturable universe has been displaced by a dizzily unimaginable one at every level. If the past sense of creation was based on the cosmic order of reality, the present poses the question anew: the order has vanished into the seeming randomness of evolution; and that past world of fixed natures has yielded to the statistical probabilities of emergence.[44]

How, then, can the question of creation be posed in such a startlingly different context? The way out of this impasse is to insist on a broadening of the experiential base of human reflection, beginning with the realization that it is *human* reflection, the activity and the manifestation of the human mind itself. Both in the past and in the present, the emphasis on objectivity is so great that the human mind is regarded as one thing in a world of other things. Thus Aquinas in a homily asserts:

> God, like a good teacher, has taken care to compose most excellent writings that we may be instructed in all perfection. 'All that is written,' says the Apostle, 'is written for our instruction'. And these writings are in two books: the book of creation and the book of holy Scriptures. In the former are so many creatures, so many excellent writings that deliver the truth without falsehood. Wherefore Aristotle, when asked whence it was that he had his admirable learning, replied, 'From things which do not know how to lie'.[45]

Here we can sense the value of objectivity arising out of a confident intellectual commitment to 'the things that do not know how to lie'. The challenge today is to reappraise the seemingly simple requirement of being objective. It consists in integrating the subjective dynamics of the mind present in the process of being objective. It is not enough to downplay the subjective as the area where lying is possible, through deceit, bias, foolishness. But a vital insight occurs when one begins to notice that any objective statement about what is the case is the complex outcome of an interior demand of self-transcendence, by focusing on what is given, questioning its meaning, weighing evidence, taking a stand on its truth, as Lonergan's axiom implies, 'objectivity is the fruit of authentic subjectivity'.[46] Subjectivity is authentic when the subject concerned is present to itself precisely as responding to all the demands of reality. To know any truth, I need to attend to all the pertinent data, to refine my sensibilities, to vitalize my imagination, to ask the unwelcome questions, to consider the possible answers in broad and differentiated fields of meaning, to ponder the emerging evidence in a disinterested commitment to truth; finally, to take my place as a trustworthy agent in the vast collaborative exercise of illuminating the manifold mystery of existence and of forming a more human world.

In genuine scientific objectivity, all such activities are present. But the typically overlooked domain of data to be considered is the data of consciousness itself, above all the consciousness of scientists themselves in their acts of understanding. It is a matter of appropriating the experience of intelligence. And even more deeply, this awareness of the dynamics of intellectual activity leads to a re-evaluation, within the experience of exploring the universe, of what was once simply called 'soul', or 'will' or 'heart'. If Aquinas considered that 'the human soul understands itself by its understanding, which is its proper act, perfectly demonstrating its power and nature'[47], a modern scientist can suggest,

> We must try once again to experience the human soul as soul, and not just a buzz of bioelectricity; the human will as will, and not just a surge of hormones; the human heart not as a fibrous sticky pump, but as the metaphoric organ of understanding.[48]

The religious significance of such a return to authentic subjectivity is expressed by Aquinas when, early in his career, he wrote, reflecting on the human soul as the most mysterious entity within creation:

> God is the greatest of all goods and more proper to each one than anything else can be because he is closer to the soul than the soul is to itself.[49]

In other words, the question of God can never be satisfactorily raised if science methodically excludes from its considerations scientists experiencing their own intelligently conscious selves. Out of that experience of intelligence and responsibility, there can arise the intimation of God, not as a patriarchal projection, nor as a temporary stopgap, nor as a particular cosmic force, not as primeval matter or universal order; but as the ecstatic intelligence and love that gives rise to the universe of knowing and responding selves. God is, thus, the Light of meaning in which the human mind participates in its progressive elucidation of what is, and in its responsible concern for what should be.

Thus, a sense of the existence of God can occur in a way in a far more intimate fashion than was the case in the classical form of the question. The former notion pivoted on the insight that no existent reality, nor all such realities taken together, was the explanation of why anything or everything actually existed. The objective inexplicable reality of the world was the starting point for proving the existence of the creator, first cause, final end, supreme perfection and so forth. The trouble with this approach is that it no longer poses the question of the creator God intimately enough. So often, the universe of reality was considered as though the human mind were some kind of neutral observer, a spectator, existing in another realm, somehow outside the whole emergent process. A poet asks the relevant question:

> but who wants to be an onlooker? Every cell of me
> has been pierced through by plunging intergalactic messages.[50]

Once you come to accept that the observing and the imagination, the meaning and the valuing are dimensions of the actual universe, indeed dimensions of creation with which we are most familiar, the question of the creation opens with a new and piercing intimacy. It is question posed from within, experienced in the very activity of knowing, intimated in the very fact that everything is in principle intelligible, and felt in the ultimate attractiveness of the values of truth, goodness, justice, compassion. In such a recovery of our knowing, wondering and responsible selves, we arrive at a sense of being beholden to an Other, the source, the enablement and fulfillment of the thrust to self-transcendence. We find ourselves, not at the controlling centre of the universe, but as participating in its limitless mystery.

You can think of Wittgenstein pondering the fact that there is anything at all; or of Einstein marvelling that the universe is intelligible; or of a Rilke recommending that poets be 'bees of invisible'; or of the poet Judith Wright feeling in her whole being a consonance with universal energies. More specifically, we can reflect on ourselves, as caught up and carried along in a great cosmic unfolding, precisely as it presents us with ourselves in this actual world, to participate in it, to celebrate its varied wonder, to contribute to its direction in love, and to yield to its promise in hope. We can be aware of ourselves ecologically, relationally located in the whole web and wave of cosmic realities that nourish body, mind and heart. All this is to occupy a present in which the world shimmers with intelligibility in our minds, thrills with value in our hearts and bursts into our imagination with its beauty. At that moment, we are not dispassionate observers, but ecstatic participants. To quote Judith Wright again,

> While I'm in my five senses
> they send me spinning
> all sounds and silences,
> all shape and color
> as thread for that weaver,
> whose web within me growing
> follows beyond my knowing

some pattern sprung from nothing—
a rhythm that dances
and is not mine.[51]

As one theologian put it,

> When the concept of the human spirit is understood ...as the mode of consciousness in which the individual feels connected to the cosmos as a whole, it becomes clear that ecological awareness is truly spiritual. Indeed the idea of the individual being linked to the cosmos is expressed in the Latin root of the word religion, *religare* ['to bind strongly'] as well as the Sanskrit *yoga* which means union.[52]

8. The God of creation

Such an experience of consciousness is the edge from which we leap into the question of creation and its creator. What is the creativity in which such evolutionary self-transcendence is made possible? What is the Light in which such intelligibility happens? What is the ultimate Goodness enabling us to be aware of such varied wonder and beauty? What is the Presence in whom all of creation is present to us, and even in us? And more deeply, for believers in the Incarnation, what is Light, the Goodness, the Presence that is conscious of itself as embodied in this universe? What is the primordial unity in which the all is a 'uni-verse', a communion, a coexistence? What is the original and attractive force that has brought forth the universe, and brought it to itself in our minds and hearts? What is the oneness, the ultimate fount and goal, that has enabled the chaos of atoms and molecules, of life forms and consciousness, of relationships and processes, to emerge into this precious instant of consciousness, of awe and responsibility, ...now?[53]

The classical way of affirming the reality of creation was to 'prove' the necessary existence of the creator. Contemplative reflection on the available world of human experience led to the conviction that nothing we directly know is ultimately self-explanatory. That comparatively simple way of philosophical contemplation yields today to controlled experiment and brilliant mathematical deduction. The enormous intricacy of the 'how' of the emergent process of the physical world seeks intelligibility in 'Theories of Everything'[54], all-embracing mathematical schemes amalgamating all the fundamental forces and particles of physics together with the structure of space and time. Compared to such, the philosophical tradition looks vague and amateurish—until one realizes that many scientists are themselves loath to allow their own science to swallow the universe whole in this way. Yet there are others content to enclose the intelligibility of the universe in some great container of mere facticity: it just happens to be.

If the question of God arises, the divinity as some ultimate factor looks suspiciously like a new version of the God of the gaps. And so we return to the point: God is not the missing factor in a mathematical or physical account of the way things are; but the all-pervasive mystery inherent in our best knowing, the Light in which the universe is luminous to itself in the human mind.

As one perceptive writer points out, historical efforts to 'prove the existence of God' were an implicit affirmation of the validity of scientific research.[55] There are interesting parallels in the experience of art.[56] The God-question is a way of insisting that there can never be a finished explanation of everything, that we exist, as it were, in an open system, opening into a limitless expanse of mystery from within our exploring intelligence. The human mind can never get to some intellectual place beyond the universe. Our explorations, be they intellectual, moral or artistic, remain within it, to converge in different ways on the original and ultimate question, posed in the relativity, the provisionality and the dynamics of our knowing. The excess of implied meaning, the drive toward the sufficient reason of all our relative sufficient reasons preclude any scientific or philosophical totalitarianism. Admittedly, the doctrine of creation can be abused by the religious obscurantist. Such did happen with Galileo, and is happening today with the more fundamentalist versions of 'Creationism'.

On the other hand, the search for the Creator arises out of the experience of anomalies in any world-picture that excludes the God-question.[57] You can take any particular actually existing entity, be it a pebble or your pet dog or yourself or scientific thinking itself. The more you explore the conditions of such success in being or acting, the more you are involved in an excess of mystery. As a contemplative searcher writes:

> Now, let's check our own experience of reality. Say you pick up a pebble. You have a thing, but you also have a horizon, as it were... When you really begin to see it, you realize it is silhouetted against something that is not seen... If you look at this pebble long enough, you will somehow experience that every 'thing' is seen against a background of 'nothing'. You always perceive thing and nothing at the same time. If you didn't perceive the nothing, you wouldn't see the thing. Now, this nothing suggests what we mean when we speak about God. God is no-thing. That is why God is not nature. God is our horizon, the horizon of no-

> thing around everything. And that no-thing is much more important to us than all the things of the world together, because no-thing is *meaning*. Meaning is not a thing; it is no-thing, nothing. And God as Source of all meaning is..., well, I am reminded of what John Cage says, 'Each something is the celebration of the nothing that supports it.'[58]

Different levels of answers deal respectively with, say, molecular structure, biochemical organization, evolutionary genus, social organization and scientific method itself. No doubt one can spend a lifetime devoted to an expertise in any one of them. But at no point are we forbidden to pose the radical question: why is there this something rather than nothing? If we are ecstatically immersed in a world of meaning, truth, value and beauty, what is the ultimate ground and goal of such ecstatic self-transcendence?

Of course, you can refuse to ask the ultimate question, and conclude that despite the intelligibility, in principle, of everything within the universe, there is no mystery of super-meaning implied. The human mind can refuse its ultimate adventure. This is not to say that the mind can comprehend the mystery implied. As Aquinas would say, we can know that God is but never what God is.[59] Indeed, we know God best when we come to realize that we do not know God: our knowledge is negative, anticipative, analogical, but never direct. What the ultimate actually is, leaves intelligence at a loss. God is not one thing among many. God plus the universe don't add up to two similar things.

With this dark affirmation of God born out of the radical questionability of the world, you don't end up knowing more about the universe, except in the sense of recognizing its emergence from an incomprehensible fullness of Be-ing, life and goodness.[60]

But this intellectual 'removal of God' from intra-mundane categories is not meant to render him absent. Rather, the reverse: God does not exist within the world as a part, or aspect, of what is. The universe exists in the boundless ocean of God's communicative Be-ing. The universe, in owing its being to Be-ing, participates in the divine reality, and thus stands forth from nothingness. Hence, to say that the cosmos or nature is God's creation, is to set the whole of our actual and potential experience of the world in a field of mystery, of super-meaning, of gift, of transcendent freedom and purpose. It makes space for the presence of creator as an all-embracing, all-originating mystery. God is 'removed' from any intra-worldly system or function: as Aquinas would say, God is not part of some genus.[61]

Such an insight also makes space for the reality of creation to be what it is, with its own laws, processes, and independent being. When there is no confusion of the creator with creation, creation is given to us to explore and the mind and heart that do the exploring are themselves dimensions of that one creation.

It follows that God as 'Be-ing that lets beings be' (Macquarrie) is more intimately present to each and every existent reality than it is to itself:

> God is present in all things, but not as part of their nature, nor as a modification of their being, but in the way something which acts is in contact with what it acts upon... Since God is by nature sheer Be-ing, it must be he who causes be-ing in creatures as his characteristic effect...God has this effect on created realities not only when they first begin to be, but as long as they are kept in being, as light is caused in the air by the sun as long as the air remains illuminated. So, for as long as anything is, God must be present to it in the way that it has being. But 'to-be' is that which is the most intimate to each thing, and what most profoundly inheres in all things: everything else about any reality is potential compared to 'to be'. So God must be in everything, and in the most interior way.[62]

Understand the 'to be' of the creature here as a process of becoming, and you have the modern form of the doctrine of continuous creation. In the words of Denis Edwards:

> Theology affirms that what enables the creature to be what it is, and enables it to become more than it is in itself, is the power of active self-transcendence, which is the pressure of the divine being acting upon creation from within... Evolutionary change is empowered by the dynamic presence of the absolute being of God. Evolutionary change occurs because of the presence of transforming love, which continually draws creation to a surprising and radically new future from within.[63]

Such a statement, owing as it does so much to the evolutionary perspective of Karl Rahner, might seem rather foreign to the classical medieval tradition, but it is not altogether so. We need not pretend that the Scholastics had any familiarity with the contemporary evolutionary model of reality. Still, there is one precious perception of the classical tradition: created autonomy does not set the creature against the creator, but is a deeper manifestation of the creative presence of God. The more autonomous and self-transcendent the created reality, the more the intimate presence of the divine is implied. In other words, the Creator truly gives being and action to creation, enabling it to act. If our modern evolutionary optic understands a self-transcending surplus or excess in the emergent process of the world, this does not diminish the creative presence of God, but manifests it more surely. As Aquinas pithily noted, 'To detract from the perfection of the creature is to detract from the perfection of the divine power.'[64]

I would suggest, too, that a sense of the mystery of creation places the current evolutionary model of reality in a gracious context. If that evolutionary model slips out of a larger context of intelligibility, it gives the impression of having an all-mastering comprehension of the emergent meaning of world. We soon begin to persuade ourselves that we actually know where we have come from and where we are going. The past and the present are emptied of any significance save in terms of endless, perhaps aimless further development. Nothing has value in itself. When a species dies out, and most do, its existence is not to be reduced simply to a failure in finding a place in an evolutionary future. Nor does the singular individual have significance only as a member of a species.

In contrast to such an ideology, the doctrine of creation places each individual existent in a world, not ultimately explicable in terms of our present understanding of evolution, but as possessing its own gifted significance in a universe of being, as participating in the mystery of Being in its own right, in a way that only the creator ultimately understands and values. In short, evolution, however much it enhances our sense of the one stream of life, however much it illumines the dynamics of development and survival, does not explain everything about the mystery of life and existence. A sense of creation appreciates the significance of all existence and all life, irrespective of our comprehension of its evolutionary value, or of its likely success. Simply to have existed is an ultimate mystery and an original gift.

Such a sense of creation reaches spiritual expression in the autobiographical report of a great Christian mystic when she wrote, 'The day of my spiritual awakening was when I saw, and knew I saw, all things in God and God in all things'.[65] A theology of creation arises from the basic overture of consciousness as adoration, thanksgiving and reverence: to be is not to be finally from, or for, or with oneself alone; it is to be ecstatic, centered in the originating mystery in which the universe is being drawn to its fulfillment. Consenting to such a universe, 'a truly wise person kneels at the feet of all creatures'.[66]

It must be noted, too, that an atheistic denial of 'God' is seldom a rejection of God in the terms we have been employing here. For the special antipathy that the non-believer feels is focused on premature and often naive religious or philosophical descriptions of God. If these suggest a smug religious answer suppressing intelligent questioning, the antipathy is intensified. When the devout mind naively proclaims God simply as a big (male) person, or as a disappointed manager of human affairs, or as the cosmic architect, or as the clockmaker who has slightly over-wound his artefacts, or as the superforce in the cosmic process, things are a little too simple.[67]

A perceptive French philosopher employs a suggestive array of images to mitigate this naive tendency to think of God in an extrinsic and static manner: 'A current of intelligibility and coherence is there; a river flows by without banks, but not without direction... A song rises which is pure music... a flux which is pure movement...'.[68] And, 'If meaning is music, the Source is not the record-player or the violin, it is rather the musical inspiration...'.[69] He goes on to reflect on the experience of meaning, and the question it poses:

> It is easier to perceive the flux of meaning than to conceive of God. One can go through life descrying the primordial harmony of the universe, without saying a thing about its origin. But is it desirable to remain there? We can understand that one might never want to isolate himself from that comprehension in which throughout the cosmos meaning discloses itself as a radiance, as an evanescent horizon, a progression, an élan. And yet we believe that it is necessary to go beyond the seductive image of a song which would be without consistency. The Source is also a subject of existence, and endowed with the extraordinary property—self-grounding. That limited, and furthermore, unimaginable, conclusion is quite certain, if one wishes to take up again, at new expense, the philosophical meditation that began two thousand five hundred years ago, and has continued ever since...[70]

In the preceding pages, we have been pursuing such a philosophical meditation, connecting it, where possible, to the God of faith, 'The One God who is above all and through all and in all' (Ep 4:6). A later chapter on the Trinity will extend these reflections.

9. Creation and the Big Bang

This philosophical and theological tradition of creation is quite compatible with the recent scientific theory of the 'Big Bang.'[71] On the one hand, it would be disastrous for any theology of creation to welcome such a scientific model as a confirmation of itself. Perhaps some version of the 'Steady State Theory' might re-emerge, or the Big Bang be located in some kind of cosmic oscillation of expansion and contraction. Who amongst theologians is to say? But what theology does emphasize is that the mystery of creation does not imply that God is a pre-temporal pyrotechnician (or pyromaniac) igniting a gigantic firecracker. For creation theology, the primary image is not a gigantic explosion as it is modelled in the domain of physical or mathematical imagination, but a theopoetic Word: 'and God said, Let there be...' It is a Word creative of this particular world, out of an infinity of divine possibilities.

If science does, in fact, continue to pursue its metaphor of the Big Bang, all theology can only say is that if God created a temporal world, then its genesis would no doubt look something like the way in which contemporary physics is describing it. But God does not have to be factored into the physical explanation as a particular, categorical cause. Rather, God is the cause operating in all causality and creativity, a 'transcendent' cause as the philosophical tradition would name it. Be-ing is not a filler of gaps, but that original matrix in which 'we live and move and have our being' (Ac 17:28), to quote Paul (probably citing Epimedines), 'above all, through all, in all' (Ep 4:6).

The creator God is not the hypothetical explanation of the Big Bang of our cosmic origins. The divine reality is sought in a much larger horizon, as the source of meaning implied, but not directly known, in the existence of a universe in which science and questions about such origins continue to be possible.

Debate in this area could be clarified by pondering the following contrasts. Creation is about adoration; the Big Bang is a plausible scientific hypothesis. Creation is about the totality of gracious mystery implied in every 'now'; the Big Bang is about a mathematically imagined 'then'. Creation is about the sheer limitless actuality that enables things to be, Pure Act; the Big Bang is more akin to the initial potentiality of a cosmic event, perhaps closer here in many respects to 'prime matter' in the language of Greek philosophy: a reality never directly known, but an essential, undifferentiated potentiality always expressing itself in new forms.[72]

Further, creation locates us in a God-given, theological genesis of the universe; the Big Bang locates us in the evolutionary genesis of a cosmos. Creation is about how everything depends on God; the Big Bang is about our common emergence from one cosmic origin. Creation narrates the story of how the eternal God acts in freedom and love to call this universe into existence; the Big Bang begins the story of the universe as a temporal and spatial cosmic reality. Creation is about an all-comprehending Providence; the Big Bang is about the singularity of a physical and mathematically computed event, and its resultant dynamic structure. Creation is about abiding mystery; the Big Bang is about the solution of a physical problem, albeit of cosmic proportions. Creation understands the meaning of the universe as a participation in the Light of God; the Big Bang has its origins in a perception of the universe bathed in cosmic radiation. Creation is, finally, a poetic whole in the sense that all being, intelligence, value and beauty are inspired by the mystery of an abiding Source; the Big Bang is an elegant mathematical formulation of the details, the proto-language, of cosmic origins.

Hence, while theology and science have plenty of opportunity to initiate a new phase of conversation on the meaning of the cosmos as it comes into human consciousness, only confusion results if various methodologies uncritically borrow from one another without appreciating the different contexts and aims of the distinct disciplines. Hence, from the religious point of view, reserve is justified in regard to Pius XII's enthusiasm for the Big Bang theory as endorsing the traditional doctrine of creation in time.[73] After all, it may be that science must be left free to investigate the possibility of a universe unfolding in phases, oscillating between a series of Big Bangs and Big Crunches. There is no point in burdening such a theory with the constraints of a religious doctrine of a creation in time, which is neither affected by, nor originally concerned with, such a scientific hypothesis. On the other hand, theology should be a little embarrassed by an astrophysicist's touching confession as he concludes his treatment of *God and the Astronomers*:

> At this moment it seems as if science will never be able to raise the curtain on the mystery of creation. For the scientist who has lived by his faith in the power of reason, the story ends like a bad dream. He has scaled the mountains of ignorance; he is about to conquer the highest peak; as he pulls himself over the final rock, he is greeted by a band of theologians who have been sitting there for centuries.[74]

Irony and rhetoric aside, such a statement hardly justifies the slightest tincture of theological smugness—if only for the reason that such a declaration could be so profitably reversed. It is more a case of theology, however secure in its religious faith, and even because of it, panting up new slopes of knowledge to find new cosmic connections in the exploration of its mysteries. To that degree, the scientists have been sitting on the peaks, waiting for theology to arrive. More sensibly, the real point is that both theology and science can be joined in the holy rivalry of humility. Neither is waiting for the other 'at the top'. Both disciplines might, however, profitably construct a shared base-camp to discuss the conditions for proceeding up the slopes as a joint expedition.

Just how much the creative Word of God will resonate with the findings of science will remain on the agenda for decades to come. Here, the venerable theological theorems of how God, the first cause, acts through all the variety of secondary causes are long overdue for revision. For divine transcendent causality does not imply that God is simply a bigger agent, but that God is God-in-action in every aspect of creation. Such action is outside all finite categories. If quantum physics has rendered obsolete naive versions of physical causality, Christian theology, for its part, has yet to explore to the fullest extent how God's presence in the world is a modality of the self-communication of God in the 'divine missions' of Word and Spirit. The divine processions within the trinitarian mystery ground and inform the universal process, to be enacted within the cosmos in the incarnation, death and resurrection of Christ. Beyond merely physical efficient causality, the causality of participation and of the relationalities of mutual love and presence must be more fully explored. The variety of 'Process Theologies' point in the right direction. But, to my mind, a more comprehensive viewpoint has to emerge.[75]

At the other extreme, the world of 'secondary causes' is now revealed as one of amazing and intricate sequences, in which a dramatic, explosive first is solidly probable. The magnitudes and potential of this first in the order of physical reality (the Big Bang) so stuns the modern imagination that theological and religious accounts of the 'first', as 'in the beginning...' feel temporarily tongue-tied. Theology will begin to be worthy of the Creative and Incarnate Logos only after a long season of dialogue with the *logoi* of cosmology and ecology, to mention the two that concern us here.

10. The experience of creation

What we have sketched here are just a few elements of the conversation that is emerging.[76] In the meantime, the poetry of creation celebrates the uncanny occurrence of all existence, as so simply and graciously given, in all its differentiation, interconnectedness, and longing for completion. It remains as a summons to participate in what is still in the making, and as a liberation from what resists the movement toward cosmic communion.

When the universe dawns within our minds and hearts as God's creation, when the divine Be-ing is recognized as the matrix from which all being emerges, as the mystery in which everything participates, our thinking—to borrow a wordplay from Heidegger—is most fittingly 'thanking-thinking'. Existence can only be thought through as a gift: not only as *data*, the given, but as *dona*, the gifts. It inspires a kind of thinking that is not centered in itself as the activity of a cosmic know-it-all. Rather, thinking is always clearing a space in the midst of a larger mystery. It speaks its words out of, and into, an original silence. Whatever clarification theology has to offer is attained only in reverence for a limitless unknown, in a more complete surrender to what is Other. Hence, the first and last movement of thought is neither control nor analysis nor solution of problems. First and last, thinking is thanking, a gratitude growing from the roots of our being, and expanding in the light of wonder and ever fresh discovery. It becomes a matter of having to give thanks to Someone for the uncanny gift of existence.

It is, too, a matter of moving into the wider world in which faith and science can meet without defensiveness or embarrassment. The following words are something of a Christian protocol of dialogue:

> We need each other to be what we must be. Science can purify religion from error and superstition; religion can purify science from idolatry and false absolutes. Each can draw the other into a wider world, a world in which both can flourish ...the vitality and significance of theology for humanity will in a profound way be reflected in its ability to incorporate these findings... The matter is urgent. Contemporary developments in science challenge theology far more deeply than did the introduction of Aristotle into Western Europe in the thirteenth century... Christians will inevitably assimilate the prevailing ideas about the world, and today these are deeply shaped by science. The only question is whether they will do this critically or unreflectively, with depth and nuance or with a shallowness that debases the Gospel and leaves us ashamed before history.[77]

11. Summary

We have focused this extended reflection on the theme of creation. I suppose the most obvious emphasis is that our thinking on such a theme happens within the Mystery of Creation, not apart from it. I have highlighted the negative character of our knowing in a way that allows for different standpoints. That point was further instanced by the way the modern creation question is occurring, somewhat in contrast to the biblical expression. From there, we proceeded to note some features of the change taking place in our perception of cosmic reality. As a resource for further reflection, we attempted a critical retrieval of some major elements in the medieval inheritance. That, in turn, led to an emphasis on the new dimension of the cosmos known to us today, and the refinement of method that was demanded. From there we went on to the central issue of the God of creation. Then, by contrasting the theology of creation theology with the Big Bang hypothesis, we found a point of humility at which both dialogue and spirituality of creation were possible.

The next connecting theme, The Human Question, will serve to emphasize in another context many of the large matters already treated, and to raise more pointedly the issue of human creativity within the world of God's creation.

1. An outstanding pioneering effort in the linking of Judaeo-Christian doctrine of creation within ecology is Jürgen Moltmann, *God in Creating: An Ecological Doctrine of Creation*, SCM Press, London, 1985. It opens out many of the perspectives we are discussing here, even though, as would be obvious, there are many points of divergence. Another work from a Protestant perspective, but closer to my own position, is Langdon Gilkey, *Maker of Heaven and Earth: The Christian Doctrine of Creation in the Light of Modern Knowledge*, Doubleday, New York, 1959: a work of enduring importance.
2. Admittedly, this more philosophical understanding does not go back to Genesis, which speaks more of a primal chaos. The *ex nihilo* character of creation was explored more in reaction to later Gnostic and Manichean teachings that supposed the a non-created evil, material principle.
3. For a broader context, see Gabriel Gomes, *Song of the Skylark I: Foundations of Experiential Religion*, University Press of America, Lanham, Maryland, 1991, pp. 171–210.
4. Bernard Lonergan, *Method in Theology*, pp. 341f.
5. See Thomas Aquinas, *Summa Theologica*, 1, 12, 13 ad 1: '...we are united to God as to one unknown'.
6. Paul Davies, *The Mind of God: Science and the Search for Ultimate Meaning*, Simon and Schuster, London, 1992.
7. Michael Polanyi, 'Faith and Reason', in *Journal of Religion* 41, 1961, p. 244. For a fuller treatment, see his *Personal Knowledge*, University of Chicago Press, Chicago, 1958.
8. Though these accounts, as we have them, were composed relatively late, probably in the fifth century B.C. For the biblical debate, see the works of Von Rad, Westermann and Anderson.
9. See Denis Carroll, 'Creation', in *The New Dictionary of Theology*, pp. 246–58.
10. Carl Sagan, *Cosmos*, Random House, New York, 1980, p. 23.
11. Jacques Monod, *Chance and Necessity*, Knopf, New York, 1970, p. 180.
12. Werner Heisenberg, 'Scientific and Religious Truths', in *Quantum Questions*, Ken Wilber (ed.), Shambhala, Boston, 1984, p. 39.
13. Though others who could be mentioned are Stephen Hawking, P. W. Atkins, Robert K. Adair, and Harald Fritsch and many others.
14. Paul Davies, *God and the New Physics*, Simon and Schuster, New York, 1983, p. ix.
15. Etienne Gilson, 'En marge d'un texte', in *Louis de Broglie, Physicien et Penseur*, Michel, Paris, 1953, p. 153.
16. As an indication of the medieval fascination with science, see Walter Principe, '"The Truth of Human Nature" according to Thomas Aquinas: Theology and Science in Interaction', in *Philosophy and the God of Abraham: Essays in Memory of James A. Weisheipl*, R. James Long (ed.), Pontifical Institute of Medieval Studies, Toronto, 1991, pp. 161–77.
17. Quoted in Michael J. Buckley, 'Religion and Science: Paul Davies and John Paul II", in *Theological Studies* 51/2, June 1990, p. 313.
18. Judith Wright, 'The Forest', in *A Human Pattern: Selected Poems*, Angus & Robertson, Sydney, 1990, p. 104.
19. Thomas Aquinas, *Summa Theologica* I, 65, 3.
20. For a constructive discussion of the role of metaphor in theology, from the past into the present, Sallie McFague, *Models of God: Theology for an Ecological, Nuclear Age,* SCM, London, 1987, is an important work.
21. Such an emphasis is, of course, no merely theological position. The statement of the Fourth Lateran Council (1215) is always a healthy reminder of limitation:

'...between Creator and creature no similitude can be expressed without implying a greater dissimilitude' (Neuner-Dupuis, #320).

22. See Arthur Peacocke, *God and the New Biology*, pp. xiv, 6. E.F. Schumacher, *A Guide for the Perplexed*, p. 5 quotes Vikor E. Frankl to the effect that 'the problem is not that specialists are specializing but that specialists are generalizing'.
23. How much so is exemplified in the Matthew Fox, *Sheer Joy: Conversations with Thomas Aquinas on Creation Spirituality*, Harper San Francisco, 1992.
24. Thomas Aquinas, *Summa Contra Gentiles*, Book 2, ch. 2. I am working from the Latin text.
25. Ibid., book 2, ch. 3.
26. Ibid., book 2, ch. 45.
27. Thomas Aquinas, *Summa Theologica* I, q. 47, 1.
28. Ibid., q. 20. a. 2.
29. Apart from an abundance of exciting documentation and marvelous instances of the variety of the past, Stephen Jay Goulding, *Wonderful Life: The Burgess Shale and the Nature of History*, Penguin, London, 1989, is a striking model of scientific reconstruction, even as it poses profound philosophical questions.
30. F. Stefano, 'The Evolutionary Categories of Juan Luis Segundo's Theology of Grace', in *Horizons* 19/1, Spring 1992, p. 9.
31. Mark Doughty, 'This Impossible Universe', in *The Tablet*, 26 September, 1981, p. 929.
32. For a fuller discussion see Stanley L. Jaki, *God and the Cosmologists*, Scottish Academic Press, 1990.
33. Ibid., p. 929. For a more philosophical approach see John Jefferson Davis, 'The design argument, cosmic "fine tuning", and the anthropic principle', in *Philosophy of Religion* 22, 1987, pp. 139–50.
34. Stephen Hawking, *A Brief History of Time: From the Big Bang to Black Holes*, Bantam Press, New York, 1988.
35. Hawking, ibid., p. 121f.
36. For some further explanation of these four forces and the efforts to bring them together in a 'Grand Unified Theory', see Ian G. Barbour, *Religion in an Age of Science*, SCM Press, London, 1990, pp. 126f.
37. David Toolan, 'Nature is a Heraclitean Fire', op.cit., pp. 18f.
38. For a fuller discussion of 'The Anthropic Principle', see Ian G. Barbour, *Religion in an Age of Science*, pp. 135–48.
39. Donald Nicholl, 'Symphony of the Universe', in *The Tablet* 16 April, 1988, p. 432.
40. For an accessible elaboration of this point, see Denis Edwards, *Made from Stardust*, Collins Dove, Melbourne, 1992.
41. As one significant effort of retrieval, see Walter Wink, *Unmasking the Powers: The Invisible Forces That Determine Human Existence*, Fortress Press, Philadelphia, 1986—especially the final chapters. for a more traditional but imaginative treatment, Louis Bouyer, *Cosmos, The Word and the Glory of God*, St Bede's Publications, Petersham, Massachusetts, 1988, pp. 194–216. Then, in a much larger ecumenical perspective, there is Bede Griffith, *A New Vision of Reality: Western Science, Eastern Mysticism and Christian Faith*, Fount, London, 1992, especially pp. 108, 199f, 236f, 268, 271, 274.
42. From an address to the Vatican-sponsored seminar commemorating the three hundredth anniversay of Newton's *Philosophiae Naturae Principia Mathematica*, Origins 18/23, 17 November, 1988, p. 376.
43. See Arthur Peacocke, *God and the New Biology*, J. M. Dent and Sons, London, 1986, pp. 21–6; 59–62.
44. Christopher Mooney, 'Theology and Science: A New Commitment to Dialogue', *Theological Studies*, 52/2, June 1992, pp. 289–330, note 55, p. 320.
45. Sermo V, in *Dom 2 de Adventu* (Vives XXIX), 194.
46. See B. Lonergan, *Method in Theology*, pp. 265; 292: 'Genuine objectivity is the fruit of authentic subjectivity. To seek and employ some alternative prop or crutch invariably leads to some measure of reductionism'.
47. Thomas Aquinas, *Summa Theologica*, 1, 88, 2 ad 3.
48. Melvin Konner, *The Tangled Wing: Biological Constraints on the Human Spirit*,

Harper and Row, New York, 1982, p. 435.
49. Thomas Aquinas, *III Sent.*, d. 29, 1, 3., 3.
50. Judith Wright, 'Connections', *A Human Pattern: Selected Poems*, Angus and Robertson, Sydney, 1990, p. 237.
51. Judith Wright, 'Five Senses', *A Human Pattern*, p. 222.
52. Robert Faricy, *All Things in Christ*, Fount, London, 1981, p. 56.
53. Denis Edwards, *Jesus and the Cosmos*, Paulist Press, Mahwah, New Jersey, 1991, pp. 44–54 uses the theology of Karl Rahner to give a concise treatment of creation from an evolutionary perspective.
54. Paul Davies, *The Mind of God*, p. 165.
55. See Herbert McCabe, *God Matters*, Geoffrey Chapman, London, 1987, pp. 2–9.
56. See George Steiner, *Real Presences*, The University of Chicago Press, Chicago, 1989.
57. Herbert McCabe, *God Matters*, p. 3.
58. David Steindl-Rast in Fritjof Capra and David Steindl-Rast with Thomas Matus, *Belongong to the Universe*, Harper San Francisco, New York, 1991, pp. 99f.
59. Thomas Aquinas, *Summa Theologica* 1, q. 12–13.
60. Kathryn Tanner, *God and Creation in Christian Theology: Tyranny or Empowerment?* Basic Blackwell, Oxford, 1988, is a most useful study of the deep grammar of the classical theological tradition.
61. Thomas Aquinas, *Summa contra Gentiles*, book 1, ch. 25.
62. Thomas Aquinas, *Summa Theologica* I, q. 8, a. 1.
63. Denis Edwards, *Jesus and the Cosmos*, pp. 53f. For a different context, see J. Moltmann, *The Way of Jesus Christ*, SCM, London, 1990, pp. 297–301.
64. Thomas Aquinas, *Summa contra Gentiles*, Book III, ch. 69.
65. Mechtild of Magdeburg, as cited in Walter Wink, *Unmasking the Powers*, Fortress, Philadelphia, p. 166.
66. Ibid.
67. For a splendid example of reworking the whole God question from a feminist perspective, see Elizabeth A. Johnson, *She Who is: The Mystery of God in a Feminist Theological Perspective*, Crossroad, New York, 1992. Unfortunately, this work appeared too late for me to incorporate its perspectives in this present treatment.
68. Olivier Rabut, *God in an Evolving Universe*, trans. W. Springer, Herder and Herder, New York, 1966, p. 11.
69. Ibid., p. 151. On music as the most suggestive experience of creation see George Steiner, *Real Presences*, p. 218.
70. O. Rabut, *God in an Evolving Universe*, p. 151.
71. For a digest of the latest scientific findings and theological reactions, Ian G. Barbour, *Religion in an Age of Science*, pp. 125–35 is excellent.
72. For an appreciation of 'prime matter' and its possible relevance to modern cosmology, see Rupert Sheldrake, *The Rebirth of Nature*, pp. 88f.
73. Pius XII, 'Modern Science and the Existence of God', *The Catholic Mind*, March 1952, pp. 182–92.
74. Robert Jastrow, *God and the Astronomers*, W.W. Norton, New York, 1978, p. 116.
75. See my *Trinity of Love: A Theology of the Christian God*, Michael Glazier, Wilmington, Delaware, 1988, pp. 189–202.
76. For further reflection, see John Polkinghorne, *Science and Creation: The Search for Understanding*, SPCK, London, 1988, especially pp. 51–68.
77. John Paul II, 'Letter to the Reverend George V. Coyne, S.J., Director of the Vatican Observatory', in *Origins* 18/23, November 1988, p. 378.

SECTION V

A Fifth Circle of Connections: Human Being

In turning now to a more explicit consideration of human existence, I begin by stating the obvious: of all the creatures in the universe, of all the variety of life-forms, the human is a question to itself. And there is no hiding from it. If the stream of life carries us on, its occasional pools of reflection are places where the deepest questions surface: where do we come from? Where are we going? How do we belong? What are we meant for? What can we trust?[1]

1. The human question

The need to question arises out of an inevitable collision of limitation and possibility. In one way, there is an obvious living unity about our existence: before any distinction or analysis, we simply *are*: 'life is fired at us point-blank' (Ortega y Gasset). Yet, as we turn to consider all the dynamics and structures implied in what and who we are, no simple description seems possible. Nature and person, body and soul, man and woman, individual and community, society and culture, cosmos and history, God and the universe—these are just some of the dualities employed in our unfinished self-descriptions. We human beings seem to be destined to make life complicated.

It is this very complication that fuels the search for some comprehensive meaning in which to focus the whole thrust of our conscious life; some ultimate value into which to pour the deepest passion to connect and belong. But how is such a meaning and such value to be expressed? With difficulty. The various levels of our consciousness, the fragmentary character of our perceptions, the diversity of our viewpoints, often enough leave us stranded in complexity.

On the other hand, we do have ourselves, or better, *are* selves, conscious in this moment, and earthed in this tiny planet in a galaxy in which one hundred billion suns are said to shine. The more comprehensively and critically we can claim the structure and dynamics of that human self in its unfolding relationality, and in its rootedness in the cosmos itself, the more telling any formulation is likely to be. The great Goethe evokes the uniqueness of human existence in his aphorism: 'Man is the first conversation that nature holds with God'.[2]

The summit of such converse has already been touched on in our reflections on the meaning of the Incarnation. In Christ, the world's path to God and God's way into the world are embodied: 'the Word became flesh and lived among us, and we have seen his glory...' (Jn 1:14). Though the Incarnate Word is replete with universal reference, there is also a sense of divine tact. Though the Word is uttered into the human history, it is not meant to stop the human conversation: '... what we will be has not yet been revealed' (1 Jn 3:2). The flow of life goes on, and history passes into new ages—as in the present when a globalization of human experience is occurring. The Word indeed was made flesh and dwelt amongst us; but that 'flesh' is a world of questions: what does it mean to hearken to such a Word in that larger comprehension of life in which the human is one of a million other species? How is the universe of current science hospitable to the grandeur and smallness of human existence? And between those two questions stands the explicitly theological one: how can Christian faith today be a conversation within the ecological and cosmically attuned culture of our day? In short, it is a good time for Christian believers to remember that the first words of the Word in John's Gospel is a question: 'What do you seek?' (Jn 1:38).

2. Language and culture

If the consideration of human existence plunges us into a world of questions, far more obviously does it put us in a world of words. We need to speak what we are, to word our experience, if we are ever to appreciate it critically. Now, the languages available to our search for human meaning often foreshorten the possibilities of an answer. For example, most of us are at least politically aware of 'sexist' language as it precludes the recognition of equality in a democratic society. If we speak as though all human beings were male, we are speaking an alienating dialect. Similarly, there is the often implicit racism of many (all?) languages. Further, the current revulsion against the flat, quantitative, purely economic description of society, so favored in modern politics, demands a more humane communication. Others note with alarm the increasing robotization of the human evidenced in such terms as stimulus and response, conditioning, input and output, turned on and switched off, being burned out or blowing fuses, doing one's thing, or being programmed or brainwashed, developing in cycles or stages; possessing 'magnetism', the right chemistry, or even the right image, and so on. There is the obvious danger of linguistically replacing the total range of consciousness with the model of the machine, of the computer, of chemical or physical interaction.

If an Inuit language has twelve words for snow, and Arabic has nine words for camel, if some Australian Aboriginal languages have some nine or ten words for water, and if seemingly primitive languages have dozens of ways of addressing the other compared to the all-purpose 'you' of modern English, you are left wondering how much we do not notice because we cannot word it. Languages can indeed blind a culture to various degrees of whiteness, or to varied features of camels, to the various qualities of water, or to the varying degrees of interpersonal intimacy. And to the mystery of the human itself.

Though the limitation of language(s) remains a concern when talking about ourselves, it is not the main problem. The real concern lies in the deep language, the culture out of which these languages emerge and in which they communicate. The more one's culture is numbed and stunted in its humanity, the more it deadens any capacity to word the whole range of human experience. It is as though there is nothing there to be said. 'Humility', 'mercy', 'sin', 'chastity', 'prayer', 'God', 'destiny' become verbal gestures to the quaint propensities of another age. And yet the words of the poet, the cadences of great music, the silence of the mystic, the radiant witness of moral achievement intimate, in their different ways, a 'nonetheless': the embarrassed or inarticulate silence can be broken. The 'narrowspeak' of our daily discourse can be refreshed with the 'wholespeak' which addresses our hearts and imaginations.[3] There is an inexpressible more, and excess, to challenge our glibness, to turn statement to question, to make our silences meaningful, to edge us more surely into the presence of that strange reality, described by each of us, as 'I am...'.[4]

If to be human is to be nourished and clothed and worded by a particular culture, what if the culture is distorted, stunted, wrong? What kind of self does it give us to live? Take this example. It is often alleged that the human has been monopolizing all our historical attention to the exclusion of the integrity of the natural world. Thus, 'anthropocentricity' is the ecological sin.

What is that saying about our culture? The following are possibilities. If a culture communicates a self-description of the human merely as an unregulated exploiter of a limitless or untamed nature, it is clearly perverted. If it defines the human individual merely according to his or her abilities to consume commodities, it becomes bloated, doomed to decompose in its own greed. Likewise, any culture that excludes any reference beyond the tiny scope of individual and usually instant experience, that speaks no language of either self-transcendence or of concern in regard to either the present or the future, may idle for a time in a tender form of self-absorption; but it moves ever closer to the psychopathic.

On the other hand, a culture which sees only the problems, with no sense of a larger grace or mercy, which is no longer able to acknowledge sin or failure, may find a fragile solace in blaming others or other times for its troubles; but that is just a step along the way to increasing depression. Perhaps in disgust, some will seek to transcend the world of problems by taking refuge in some form of mystical otherworldliness. But schizophrenia is of limited creative value. Further, any way of life that begins to prize the image over reality soon finds everyone wearing masks in a wilderness of mirrors. Finally, any culture that systematically represses any type of experience, that prevents certain questions being raised, that reduces all values to personal preferences, is on the way to becoming a slum.

Now, if 'anthropocentricity' means any or all of these cultural diseases, then it is clearly destructive. Such cultural distortions are what Christian tradition knows as 'original sin'—that mutilated sense of self communicated by the culture in which one lives.[5] The following quotation makes the point:

> The socio-cultural character of original sin is thus evident. If original sin means treating as non-existent the dependence of the sense of man's meaning and value on the ultimate mystery, it is in society as a whole, where the human sense of man's meaning and value are stored, expressed in institutions and in the whole cultural achievement, that we shall expect to find this non-acknowledgment, this conspiratorial silence as to the ultimate source of man's sense of meaning and value. Original sin is the universal and socialized withdrawal of man from the mystery on which he continues to draw his meaning and value. Original sin is the socialized truncation of human life, the systematic reduction of the child of mystery to the banal world of man's own making. [6]

What, then, is the path forward? It has to be by way of reclaiming an authentic sense of the human self. The more that larger consciousness of the self remains disowned, the more it remains mutilated, stunted, or perverted by cultural bias, the more elusive any liberation appears. Take the ecological crisis itself. We cannot hide from our intelligent and responsible humanity. We have to face up to what we have made of ourselves in our history, especially in the last two hundred years or so. It is in our minds and hearts that the ecological issue has to shape itself into a world-transforming option.

Is that overemphasizing the human? Perhaps. Still, it is not customary for leopards to take courses on how to change their spots. Kangaroos do not hold seminars on population control in a drought-stricken national park. Seals do not research reports on the problems of fish breeding. Neither the AIDS virus nor malarial mosquitoes are notably suggestible to the desirability of a larger point of view. No doubt, we humans have much to learn from what is perceived as the wisdom of dolphins or the tribal bonding of chimpanzees. But they do not seem weighed down with our precise problem. The Disney-fication of the animal world is a lovely fantasy, but a limited moral resource.

Where then does that leave us? For better or worse, we are stuck with ourselves; and with our uniquely responsible place in the world. We cannot hide. It is from within ourselves, and from within the culture that embodies our meanings and values, that new energies and direction must emerge. If there is to be any escape, any release into a new belonging, it must be through facing what we are, and what we have in fact chosen, one way or another, to become.

3. 'Anthropocentricity'?

Hence, ecology, in the first place, is a conversation conducted by human beings; by men and women; by the old who remember other times, by the young who dream of new times; by the artist, the religious believer, the scientist—all for the purpose of inspiring a new human connection with the whole. It may be about many other forms of life, but it is not addressed to them. It aims to get a hearing for 'the whole' in the chattering 'narrowspeak' of most of our daily or educated discourse. A philosophically minded ecologist writes:

> A humanity that has been rendered oblivious to its own responsibility to evolution—a responsibility that brings reason and the human spirit to evolutionary development, diversity and ecological guidance, such that the accidental, the hurtful and the fortuitous in the natural world are diminished—is a humanity that *betrays its own evolutionary heritage*. It surrenders its species' distinctiveness and its uniqueness. It is grossly misleading to invoke 'biocentricity', 'natural law', and antihumanism for ends that deny what is distinctive in all human natural attributes.[7]

The issue is not that of downplaying or dismissing the importance of the human, but persuading the human to be more involved with its obvious reality. More deeply, it is a matter of making a clearing, as it were, in which the wholeness of reality, in all its splendor and connectedness, can emerge within the human mind and heart. Bookchin goes on:

> Viewed heuristically as an effort to bridle 'human' arrogance toward other life-forms, one can sympathize with a 'biocentrism' that seeks to provide a counterpoint to the present society's destruction of the biosphere. How long one can continue to belabor 'Humanity' for its affronts to the biosphere without distinguishing between rich and poor, men and women, whites and people of color, exploiters and exploited, is a nagging problem that many ecological philosophers have yet to resolve, or perhaps even recognize. 'Anti-humanism', for all the caveats that qualify it, strikes me as bluntly misanthropic and sounds less like an ecological principle than an argument against the human species as a whole. 'Natural law' theory, conceived not as a radical theory of 'inalienable human rights' but as the expression of unrelenting necessitarian nature, is riddled by troubling ambiguities.[8]

Any ecological thinking that has as its model the disaffected, solitary researcher conversing with animals or communing with nature is off to a bad start. Self-hatred provides little promise of a better world. To hear Bookchin again:

> The concept of humanity (which seems to be the one species that has the ability to luxuriate in self-hatred) is no less sucked into this ideological black hole. This phenomenon has several names: a logically ambiguous 'biocentrism' and often strident 'anti-humanism' that are set against 'anthropocentrism' and 'humanism'—presumably the two cardinal sins of abstract 'Man' who is determined to despoil an equally abstract 'Nature'. Just as non-human life is divested of its specificity, so human life is divested of social development. 'Society', too, becomes an abstraction that somehow leaps into existence without any relationships to nature, not to speak of such compelling characteristics as hierarchy, domination, class conflicts, and the emergence of the State. Out of these abstractions emerges a simplistic biologism, often structured around 'natural laws', that sees 'Man' and 'Humanism' as a curse that afflicts 'Nature' with ecological degradation.[9]

Hence some kind of relational anthropology is inescapable. If I deny the critical centrality of the human, at what center do I stand? In God? Such a standpoint might promise a splendid advantage, but it is apt to become onerous in the responsibility of upholding the universe. In 'Nature' or 'life'? However alluring such standpoints might be, they logically locate one outside the human realm. We begin to imagine ourselves as reduced to the purely animal in the denial of any cultural capacities, or as elevated to a psychotic superconsciousness that can bypass human learning and accountability. So, whatever the enmity or the outrage in regard to 'our own', there is no denying that it is with 'our own' that we must begin. Neither a lower or higher eccentricity is likely to have much influence on the historical decisions *we* have to make in a world that is not Disneyland. Nature is not inhabited by animated cartoons with dubbed human voices. However endearing this might be on a certain level of fantasy, it would be as unwise to stop there as it would be unrealistic to model justice after *Trial by Jury* or *The Mikado.*

4. The relationally human

In short, we cannot escape from ourselves merely by decrying anthropocentricity. To criticize the psychopathology of a culture is one thing; to occupy a completely eccentric, non-human standpoint is another. It is, once more, a matter of starting from what, who, where and when we human beings are. To that degree, the human is the eco-subject, the freedom from which decisions are made to either respect or deny the integrity of the ecology of our planet:

> To recover human nature is to 'renature' it, to restore its continuity with the creative process of evolution, its freedom and participation in that evolution conceived as a realm of incipient freedom and as a participatory process.[10]

In this perspective, the human is not just one more species, if only for the reason that we are the species which names other living things in the ongoing history of science and skill (Gn 2:19–20). In human culture, the world comes forth, to be celebrated and cared for, shaped or used, preserved or transformed as the case might be. Humanity is the species that can think and choose ecologically, the species in whom the fate of the bio-cosmos can be decided.

All this is to say that the 'anthropocentric' bugbear of much eco-theological writing has, then, to be more carefully considered than it normally is. The human is not a center in which we end, but a center from which we begin; not self-centered fixation, but the self-transcendence into the other, the more, the whole. It is consciousness as inclusive relationship. Indeed, the human is nature become conscious of itself as an open system of meaning and freedom. If a naive ecological concern too often envisages itself existing in self-contained pure spirit benignly haunting an unspoiled wilderness, the critical reappraisal of the meaning of the human not only makes ecology more realistic; it also makes humanity more responsible.

It has been alleged that Christianity is anthropocentric in its concern for a transcendent human salvation; and thus, to the forgetfulness of the rest of creation. But that is too simple, at least in the present stage of discussion.[11] For Christianity faces its problems in the same way that sound ecology does. It may start with the human problem—the problem of despair, isolation, alienation from oneself, from others, from creation, from God. The release that the Gospel promises, as grace replaces sin, as the heart of stone yields to the heart of flesh, is in relationships, restored, rendered creative in liberating love, and operating out of a universal hope. That is not to be fixated in an isolated standpoint. It means, rather, an expansive existence unfolding after the pattern of an open circle in a growing circumference of concern. Notably in the present age of history, and not without the impetus of earlier forms of ecological protest, this concern has come to include the integrity of the biosphere of this planet as it is set in the increasing wonder of the scientifically explored universe. Such a movement is not centripetal, but centrifugal. It opens out to the all, the whole, the universe. It is a lived relationality, drawing its energy and form from the relational life of the trinitarian mystery itself.

Let us now approach more directly the meaning of 'human being'.

5. A classical definition?

The ancients described the human as *zoon logikon*, *animal rationale*, literally, 'the thinking-speaking animal'. They understood that there would be no 'animal' unless there was a healthy *vegetale* component and the right proportion of *minerale*. Indeed, for them, 'proportion' and 'component' are not exactly the right words. In fact, we find an instructive dispute amongst the philosophers and theologians of our premodern times on how the 'souls' of other forms of mineral, vegetal and animal existence were contained in the human soul.[12] What they all agreed on was that the human spirit essentially contained these dimensions in its constitution. Aquinas could say, out of his conviction that the human self was essentially embodied, *anima mea non est ego*: 'My soul is not the [whole] "I"'.[13] The human self was more than the purely spiritual. It was understood to be truly embodied in the earth in a way far more intimate than modern imagination or philosophy usually credits.

True, such a conception of the human concentrated on the *rationale*, the uniqueness of the spiritual dimension, where intellect and will were the supreme powers of the distinctively human 'soul'. A philosopher such as Aristotle could say that 'the soul was in some way all things': what is distinctively human is the potential outreach to everything in knowledge and love. The human was the eminently relational. Understandably, this expansive relationality was the special domain of theology. Not only did we human beings have a relationship to all that is within the universe, but to the all-originating principle that is the beginning and end of all that is, was, or could be.

Though such a synthesis had its own classic power in emphasizing that the macrocosm was somehow embodied in the human microcosm, the history of thought shows frequent distortions. Human reality was no neatly defined 'nature' in this philosophical sense. Spiritual writers generally reverted to a Platonic understanding of human existence. They dealt primarily in terms of sense-bound passions and worldly distractions which they identified with the *animal* genus of our existence, and the true spirituality which they associated with the specific *rationale* of our nature. As the essential embodiment of human existence was downplayed, the specifically human was conceived of in a more and more immaterial a fashion: the truly human drew closer to the domain of 'pure spirits', to feel itself exiled on the earth while awaiting an ever-deferred spiritual fulfillment in heaven. The pluriform nature of human existence was slowly reduced to an awkward dualism.

Today, Aristotelian-Thomist anthropology has, at best, retired to the status of a classic source, as we look for a more holistic self-definition. How much has changed is evident in the perceived quaintness of the classical terms once employed in defining the human: 'animal' has become something of an insult. And 'rational', having lost the contemplative expanse of its original meaning, is now heard almost as a taunt, implying either a desiccated logic, or an intellectuality that has forgotten how to feel for anything not contained in a laboratory. Such terms jangle in modern sensibility as shards of broken meanings.

In the transition from classical to an empirically conceived notion of culture, philosophy has yielded to experiment and observation. The universe of definable 'natures', expressed in classical languages and in the terms of a perennial philosophy, has had to come to terms with the human as a manifold *datum*, in its varied cultural contexts, and within an evolutionary world.

In such a world, 'in which everything has changed except our way of thinking' (Einstein), we need to be far more sensitive to the broader issues of culture in which humanity is defined. Typically, it seeks to incorporate a new sense of embodiment in the cosmic, biological, social process of history. At the same time, it is a search for a new spirituality, in which the irreducible specifics of human consciousness will be respected and enlarged. More particularly, the challenge is one of bringing matter and spirit into a new human unity. Consciousness cannot be left schizophrenically unrelated to the world, just as that world cannot be considered shut against the most mysterious of all phenomena, human consciousness itself. Dualism, the house divided against itself, is recognized as the danger to be avoided.

The modern discussion has shifted from a philosophical definition of the human to a far more concrete exploration. Human culture in all its actual varieties, and human history as the theatre of human self-determination, are at the forefront of any reflection on the human. The notion of a determined, universal 'human nature' has receded into the background, as a question indefinitely deferred until all the data are in. It can only become a question within the meanings and values that inform given cultures. It only arises as a question when our accelerated history makes time to consider it.

What is most obvious is that culture and history are expressions of what has been called 'man's making of man', of a profound or shallow, or a broad or narrow self-expression. This self-making is embodied in the symbols, the language, the art and the institutions, the technology and the inherited traditions structuring human communication. Each of these elements, in its own way, objectifies and transmits a certain sense of the human in terms of what life is, or might be, all about.

Now every culture expresses a peculiar specialization. For example, Australian Aboriginal culture, while it may well be classified as primitive in terms of technological prowess, is now newly appreciated as far wiser than it was possible to consider in the intellectual arrogance of Enlightenment colonization. Its capacities for long-term survival are freshly valued. For we now live with the embarrassed realization that the myths of mere progress offer no real future.

Part of the present problem is the intense irritability modern men and women feel, plunged as we are in often extreme and always competing human expressions that constitute contemporary pluralism: the 'multiculturalism' of cosmopolitan democracies, and the startling variety of differing lifestyles. If society is not going to be subject to progressive disintegration, some kind of fundamental commonality has to be achieved. The alternative is the endless dialectic of point-scoring ideologues, as the various 'silent majorities' watch by, disaffected from the whole cultural or political process, as when the best lack all conviction while the worst are filled with passionate intensity, if the poet is right.

6. A new science of the human

The only recourse is to come to new self-possession. How to own, or even 'own up', to our existence in a more inclusive and imaginative way, both as embodied in the cosmos, and in the indefinability of our human being, is the issue.

It is only comparatively recently that science has begun to pay specific attention to the human. The peculiar complexity of the human phenomenon was lost in other considerations. But the hundred organs, the two hundred bones, the six hundred muscles, the billions of cells, the trillions of atoms which somehow conspire to constitute our embodied selves are a phenomenon calling for far more extensive exploration. For scientific exploration has moved with exquisite sophistication into the very large and very small dimensions of reality. The quasar and the quark are more easily named than the human self. Science becomes more at home in the intergalactic than in the interpersonal. It is more familiar with sub-atomic indeterminacy than with human freedom and human destiny. In comparison with the human phenomenon, the entities investigated were relatively simple.

Though chemistry and biology have dealt with the molecular and genetic structures of living things, they have paused, comparatively tongue-tied before the complexity of the human phenomenon. It has been easier for the scientist, as for us all, to avoid eye contact with another human being or to ask what mystery has awoken in the consciousness we have of our selves and others. It was perhaps safer to leave the human simply classified as one of the millions of species on this planet.[14]

Now, as the physicist Heinz Pagels has noted, matters have begun to change. The study of the complex human phenomenon, in which material conditions support human consciousness as that consciousness illumines and directs nature with its distinctive transcendence, will be 'the primary intellectual challenge to our civilization for the next several centuries'.[15]

Might I emphasize once more the key issue. Because the human is so physically insignificant in the known cosmos, because of the old scientific fiction of totally objective observation, scientists seldom notice that the operation of their own minds is the most refined and complex reality the physical world has yet produced. In charting the size of the universe or the buzzing intricacies of the subatomic world, in tracing the organic mechanisms of life, or in considering the genetic interrelationships existing between all living things, the human mind catches itself in an astonishing act: in knowing and exploring the universe, it is part of that universe. And the universe, so known and explored, is becoming progressively luminous to itself in human consciousness.

Science has been so typically taken up in investigating non-conscious objects, usually as they emerge out of our past, that it tends to be distracted from the luminous occurrence of the present. But in this elusive, unobjectifiable *now*, scientific intelligence, to say nothing of artistic inspiration or mystical experience or ethical action, is in operation. Once more the lament: scientists too often leave themselves out of their own considerations! As one writer put it, 'If Carl Sagan is referring only to the exteriority of nature when he insists that the cosmos is all there is, he is clearly wrong: there is also Carl Sagan looking at the cosmos and trying to make sense of it.'[16] The most significant happening in the universe is going on behind the eyes of the observer: for in the one and a half kilograms of brain matter, it is estimated that there is more operational complexity than in the whole of the Andromeda galaxy. There are a trillion neurones of the human brain, each cell of which communicates with at least a thousand others. That would mean that the number of possible associations might exceed the number of atoms in the known universe.[17]

We are under the necessity of respecting a kind of twofold movement. The Aristotelian dictum 'causes can affect one another, but on different levels' (*Causae sunt ad invicem causae sed in diverso ordine*) is looking to a larger application. Scientific exploration has been largely concentrated in determining causality 'from below upward'. The physical, chemical, botanical, zoological are progressive approximations of the manner in which human consciousness has emerged on this planet. This movement poses its own questions: where does this emerging 'we' go from here? What are we to do in, and with, nature come to consciousness in ourselves? What are we to do with our souls?

But there is also the comparatively neglected movement 'from above downward'. Human consciousness, in the transcendent fulfillment of faith and love, in its scientific, artistic and ethical capacities, in the whole cultural structure of its history, is at the point of making a fresh appropriation of its cosmic and biological origins. What we are as human is caused by what has gone before, but also can now influence what it has emerged from. This, too, poses its range of questions: what value do we need to put on this process of emergence? What responsibilities of care now engage us to respect the physical, chemical, biological values that underpin our well-being and cultural development? How are we to care for the earth as the shared body of our co-existence? It is, thus, a matter of reflecting far more deeply on the hylomorphic inheritance of philosophy, to ask in a wider range of reference, how is matter looking to progressive animations by the spirit? How is the human spirit natively embodied in matter?[18]

The soul-values that motivate self-transcendence—meaning, truth, morality, religious faith—if they are to flourish, must be now reconnected with the 'material' values out of which human life has emerged. It has become a matter of promoting a more conscious embodiment. The human self has to be reimmersed in the physical, chemical, botanical, zoological processes that characterize the emergence of our world.

We read in the life of Berlioz that one day on a visit to Rome, he fell into the Tiber. He surfaced, singing a refrain that had long eluded him.[19] The dripping composer is an image of our present challenge: to fall into the stream of physical reality to recover the full symphonic sense of our existence.

On the subject of the symphonic sense of human being, George Steiner's remarks on music are deserving of mention:

> No epistemology, no philosophy of art can lay claim to inclusiveness if it has nothing to teach us about the nature and meaning of music. Claude Lévi-Strauss's affirmation that 'the invention of melody is the supreme mystery of man' seems to me of sober evidence. The truths, the necessities of ordered feeling in the musical experience are not irrational; but they are irreducible to reason or pragmatic reckoning. This irreducibility is the spring of my argument. It may well be that man is man, and that man 'borders on' limitations of a peculiar and open 'otherness', because he can produce and be possessed by music.[20]

Steiner cites the words of a Renaissance philosopher of music, Gioseffo Zarlino, in his conviction that music 'mingles the incorporeal energy of reason with the body'; and Schopenhauer: 'music exhibits itself as the metaphysical to everything physical in the world... We might therefore just as well call the world embodied music as embodied will.' Music incorporates a sense of reality as 'promise', the experience of the unfulfilled as both a challenge and a consolation.[21]

While I am in no position to take this insight further, it does tease us into a more symphonic appreciation of our being in the world. More generally, we are being invited into a more disciplined attention to the embodied consciousness that each of us is—the *somebody* that is existing here and now. Neither you nor I are an inert lump of matter, nor a pure spiritual consciousness cut off from material conditions, but a strange interrelationship between the two, as in the notes of a musical scale, a play of harmony, an elusive melody as a variation in a larger symphony.

For instance, we cannot think unless through some intimate connection with what our physical senses have seen, heard, tasted, felt, smelt. Even for such preliminary sense knowledge to be possible presupposes a million physical, chemical, biological, and anatomical systems to be smoothly functioning in us. To that degree, we are a system of systems, a form of existence subsuming many other interrelated and interconnected forms of being. But as we touch, taste, see, hear, imagine, think, love, work, pray, and express ourselves in work and artistic creation, we are a living whole. We are each a self in the process of becoming, an identity achieved in going beyond itself in relation to the all, the whole, the other and the ultimate. The axiom of ancient philosophers that spoke of the soul being somehow 'all things' in its capacity to know, has now to be complemented by the incorporation of our bodily being in the 'all things' of the cosmos itself. Teilhard de Chardin perceptively remarked:

> My own body is not these or those cells which belong exclusively to me. It is what, in these cells and in the rest of the world, feels my influence and reacts against me. *My* matter is not a *part* of the universe that I possess *totaliter*. It is the totality of the universe that I possess *partialiter*. [22]

All this is to say that complex levels of interconnectedness fuse and focus in the concreteness of human being. If we are to cultivate a new inclusive sense of human integrity, it will be literarily 'of vital concern' to recognize this symphonic complexity.

7. Self-transcendence in an emerging universe

Here a basic model for our self-exploration can be suggested: self-transcendence. Though this fundamental self-description is not primarily a philosophical category or an element in scientific theory, it is an integrating image occurring 'in action'; in the immediacy and momentum of actual living; in actual flow and expansion of consciousness that constitutes our communicative being in the world. A process of becoming, of going out of the self, in the direction of what is other and more characterizes authentic self-realization. Conscious 'self-transcendence' is the structured momentum of life as we instance it.[23]

Our senses place us in an environment. Our intelligence awakens to a field of forms and meaning. Our capacities to reflect and judge make us trustworthy mediators of the real world. Our feelings for the good, and our decisions in its regard, invite us into a collaborative world of moral value. For people of faith, God's gift of love places them in a universe of ultimate meaning and worth.

Regrettably, the dynamism of self-transcendence is not infallibly realized. We can be distracted from the attentiveness our senses make possible. We can avoid the questions that intelligence suggests. We can shut out the weight of evidence. We can prefer self-gratification to moral responsibility. We can opt for self-enclosure against the self-surrender that faith inspires. Still, the invitation is insistent: to go beyond self-imposed isolation into participation in a universal event. We ignore the summons into that world of meaning, value and ultimate grace at the peril of denying our true selfhood. Of this possibility, an uneasy conscience is the most intimate witness.

The way into the movement of universal becoming can only be that of consenting to dynamics written into our conscious being. While the possibilities of self-transcendence are experienced as an appealing summons, its dynamics are accessible only through discerning self-appropriation. Such self-owning is the continuing personal experiment involved in authentic existence. Now, all experiments presuppose data of one kind or another. In this instance, the data on the direction and promise of life are found in human consciousness itself. More concretely, in the way we are, and act, and love and search. Already we are living selves ecstatic in regard to a larger whole; emergent out of an immense process; beholden to values; communicative in a world of meaning, truth, value. Teilhard de Chardin sets this ecstatic emergence of the human in the widest possible context:

> Man is not the center of the universe as once we thought in our simplicity but something far more wonderful—the arrow pointing the way to the final unification of the world in terms of life.[24]

To be human is to be progressively immersed in a universal process of becoming and communion. Now, it is possible to understand so much of contemporary cosmology and natural sciences as suggesting both the pervasive pre-personal structure underpinning distinctively human self-transcendence, and as beginning to outline the emergent universe in which such transcendence is realized. The evolutionary paradigm discloses an upward pull, a 'vertical finality', a 'genetic throbbing' of the universe in the direction of life and consciousness.[25] The 'non-' of the new sciences and methods—'non-equilibrium' principles, 'non-linear' dynamics, 'non-entropic' structures—point to a search for a vocabulary appropriate to the wonder of universe as it emerges in contemporary investigation. The range of related sciences and methods is of great complexity and sophistication: non-equilibrium thermodynamics, synergetics, hierarchy theory, autopoetics, general systems theory, fractal geometry, catastrophe theory, cellular automata, non-linear chaotic dynamics. What is remarkable is how these new kinds of exploration converge. They present the emergent cosmos as a single wave, gathering into unique intensity in the human, in an ocean of undular being. Human consciousness crests in a 'waving universe'.[26]

So great a consonance is there between human consciousness and the matter of the cosmos that one scientific writer, Danah Zohar, in her *The Quantum Self*[27] has used analogies drawn from the subatomic quantum realm to further an understanding of the human consciousness in its individuality, creativity and manifold relatedness. Doubtless, many aspects of such an imaginative project will remain arguable. The valuable point of such efforts consists in linking human consciousness in some way to the intricate dynamics of a mysteriously relational quantum world. It shows that we are coming to recognize our elemental cosmic location in ways that have never been thought of before.

The point of such efforts, anthropomorphism aside, is actually to locate the distinctiveness of the human in the subatomic dynamics that quantum physics is exploring. Our immersion in the physical processes of the universe is thus recognized as a human value. Such is an index of the radically new self-appropriation offered to us as we find the fundamental movement of our being inscribed in the most elemental dynamics of matter.[28]

Still, there is an even larger complexity. Levels of living and non-living things exist below and before human consciousness in a hierarchical scale. They are related to one another not as closed but as open systems.[29] Whilst each level has its own laws and schemes of recurrence, there are instances of randomness, openness, 'chaos' even, as entropy is dissipated, and possibilities of higher realization occur: each level can go beyond itself in its own kind of self-transcendence. Subatomic particles assemble themselves into the elements of the periodic table. Chemical elements come together into increasingly complex compounds. Such compounds structure enzymes and acids to form the living cell. Such cells come together in all the variety of plant and animal life. These, in turn, are subsumed into the astonishing emergence of the human brain in the human body, just as they enter into the ecological system of the whole human life in the food we eat, the air we breathe, the water we drink, the wine of our celebrations...[30]

Now this is far from a world of fixed and static natures more or less blended or juxtaposed. What emerges is more a kind of participatory interlinking of all levels of reality in our constitution.[31] Human consciousness becomes cosmic. The universe is conscious of itself in the human body, mind and heart.

Hence, the appropriation of the dynamic datum of human consciousness cannot ignore the laws of physics, chemistry and molecular biology written into our emergence. Though theology concentrates on the ultimate self-transcendence of the human person in regard to God, it cannot disregard the activity of phenomena on prior levels, in the complex relationality of atoms, neurones and DNA molecules, cells, organs, organisms, bodies, bondings, populations, ecosystems... At each step, a higher possibility is preformed. Without such successive integrations, there would be no higher self-realization. Arthur Peacocke gives a useful summary:

> There is a continuous spectrum of levels in the total human unit, and these need to be addressed by language appropriate to the level in question. The reason all these languages, whether scientific or theological, ought to be able to communicate with each

other is precisely their reference to this objective reality and unity of human beings, to which science and theology bear witness.[32]

8. Seven levels of the human good

I will now list some seven levels that dynamically structure human existence. Unfortunately, a pitifully schematic exercise such as this leaves so much out. Indeed, a number of further divisions and subdivisions should ideally be added. On the other hand, such a list is sufficient to generate libraries of questions relevant to this historical phase of exploring what it means to be human.[33]

i. The physical level

In an ascending order there are, first of all, the physical elements and dynamics of the cosmos.

In accepting the physical 'forebears' of our existence, we are defining ourselves into the fifteen-billion-year-old story of the emergence of the universe. Whatever the ideological differences that divide the human community, that enormous expanse of time, with all its blazing energies, is our 'common era'. Our common clay is now stardust—whether we be human, mineral, plant or animal. A great inclusive 'we' extends through an indescribably vast, energy-laden flame of cosmic emergence. The whole unfolds as an all-inclusive, self-organizing process. Our shared present is implicit in the initial fireball of our cosmic origins, as it unfurls into a billion galaxies, and condenses into the elements which at this moment are firing the energies of our brains and the beating of our hearts, in all our happy capacities to wonder and to hope.

Science has sharpened its vision to detect the light that irradiates everything. The COBE satellite has given human eyes the sight to peer back fifteen billion years to their cosmic beginnings. We are the first generation to have that kind of empirical view of the origin of the universe, thereby to locate ourselves in the universal process.

Such is the prelude to any appreciation of the amazing creativity of our own planet earth. To look back those four and a half billion years is to witness a vast cauldron of activity in which the elemental chemicals were brewed. As such elemental matter slowly cools to crystallize and condense, it combines and complexifies until that point where primitive life emerges in the oceans. Out of these oceans come, in their time, creatures of the shore, of estuaries and marshes, to spread eventually over dry land, to take root in the soil, to float or fly in the air, to move on paw or hoof or foot in their various habitats, and thus to occupy the planet. Finally, there comes the human as 'the latest, the most recent extravagance of this stupendously creative earth'.[34]

ii. The chemical level

Implied in the above remarks is the necessity of owning also the chemical values of our existence. Phosphorus formed in the heart of the stars gives us the skeletons that structure our bodies. The stellar iron enters our blood. The sodium and potassium that drive signals along our nerves are part of a larger message. The cosmic flame of hydrogen burns in our brains. Carbon molecules fuel our metabolism. Our lungs breathe an inspired past.

But if the chemical balance so patiently achieved in the emergence of our planet is skewed by violent interruption, water is polluted, the air begins to carry poisons, the sunlight is dimmed by smog, all life suffers, to become blighted in its fundamental systems. The one hundred and sixty billion tons of carbon spewed into the atmosphere by modern industry over the last century and a half has made a difference, registered in the temperature of the earth. The boon of modern refrigeration has become a mixed blessing as chlorofluorocarbons begin to change the nature of the atmosphere itself, even the quality of the sunlight, as the ozone layer is ominously holed. As the monocrops of modern agriculture cover huge farming areas, the precious topsoils that took millions of years to form are eroded, to be blown or washed away.

To be human is to be chemically linked to the earth and the cosmos itself. Earthy questions arise: the most obvious bear on the nature of the food which the earth offers for our sustenance—how is this affected by artificial additives, by the Fast Food industries of our day? For some forty years, problems associated with pesticides and food additives have been increasingly recognized. Carcinogenic has become, literally, a household word and the easy availability of 'junk food' continues to be a sign of progress! But so deeply are chemically questionable practices embedded in the huge agribusinesses and 'food chains' that structure so much of the world's production, transportation, preparation and marketing of food, that the cure of ill-effects is an easier matter to address than prevention.

The issue, along with so many chemical distortions of organic life, as with synthetic hormones and addictive drugs, is how long such experimentation with human well-being can be chemically tolerated? A reclamation of our earthly chemistry is the only way to avoid long-term degradation in our bodily and psychological make-up.

Then there is the question of waste disposal and how it affects the chemical integrity of the land and waters: earth cannot sustain a limitless number of flush-toilets! The *diabolic* project to live apart from nature has to be replaced by a *metabolic* location with and within nature.

Thomas Berry, C.P., gives an incisive summary of the situation:

> The issue now is of a much greater order or magnitude, for we have changed in a deleterious manner not simply the structure and functioning of human society: we have changed the very chemistry of the planet, we have altered the biosystems, we have changed the topography and even the geological structure of the planet, structures and functions that have taken hundreds of millions and even billions of years to bring into existence. Such an order of change in its nature and in its order of magnitude has never before entered either into earth history or into human consciousness.[35]

iii. The botanical level

Much of what has been said above is relevant here, in terms of the quality of the food, soil and air of our planetary life support system. But two further considerations are suggested, as, at the botanical level, we enter the domain of living things with whom we are companions in a planetary symbiosis. First, it is a world of beauty. Here trees tower above us and blossoms fill the air with fragrance. Here flowers delight the eye and great forests are wonderfully alive. To be without them is not humanly imaginable, but in vast areas of the planet, both vegetation and the human spirit are wilting under the onslaught of soulless exploitation. In the beginning, we were meant to live in a garden...

Brendan Lovett expresses the second point:

> Further important insight gained into the human good at this level relates to the diversity of plant species. The minor issue here is the destruction/loss of local varieties of seed in favour of high-yield hybrid varieties. The latter are notoriously vulnerable and do not breed true. The major issue is the wipe-out of millions of years of nature's crucial survival experiments under the most demanding climatic conditions that is involved in the annihilation of the tropical rain forests. We are destroying the gene-pool of successful wild varieties of plants at exactly the same time as we are placing the world at risk, foodwise, through reliance on climatically highly vulnerable hybrid varieties.[36]

That puts it well. We are faced with a strange choice: a living, breathing world of healthy variety; or something far inferior.[37]

Questions do arise: how much interference can our planetary well-being and health tolerate? How much risk should we incur in modifying biochemical structures? What is happening to the basic 'food chain' linking all living things? 'The Great Chain of Being', beloved of philosophical tradition, corrodes in its biological links if the 'food chain' is polluted. How can we human beings live less diabolically and more metabolically with the earth? How can we celebrate the eucharist as the body and blood of Christ made present in the shared 'fruit of the earth and work of human hands' if the bread and wine are contaminated?

iv. The zoological level

To be human, on this level, is to be 'animal' with the animals. Here again there is not only a symbiosis, but a companionship, with the wild and the tame. We befriend nature in our pets, contemplate it in the wild, harvest it for our food, harness it for our purposes. We are the species that names the animals, and, ideally, constructs the ark for their preservation, irrespective of their immediate usefulness.

Here I will make a longer remark, and a surprising one, at least to the modern mind, long oblivious of the classical *animal rationale* specification of the human. One point is emerging from sociobiological sciences: there is much to be learned and reclaimed in human behavior by setting ourselves in the evolutionary emergence of the animal. It is a matter of recovering the animal in us if we are more adequately to understand ourselves. The details are subject, of course, to wide-ranging debate. But balanced perceptions are beginning to emerge. We are not 'disembodied intelligences tentatively considering possible incarnations' but concrete embodied human beings with 'highly particular, sharply limited needs and possibilities'[38] Our capacities to bond, to care for our young, to feel for the whole group are rooted in our evolutionary animal nature: 'We are not just like animals; we *are* animals'.[39]

Such perceptions counteract the prevailing liberal conception of the human person as an isolated, self-sufficient individual, jealously asserting his or her own rights. It also presents a healthy challenge to any theology of universal love which might tend to be excessively ethereal and unaware of inherent limitations. Our animal nature presents us with a priority of feelings and actual bondings that demand to be respected. For Thomistic theology with its doctrine of the *ordo caritatis* (the order of preferential love),[40] there is nothing odd about such a conception. However, the drawback in Aquinas's account of the integration of natural relationships into supernatural love is the Aristotelian biology on which it was founded. The most notable deficiency in such an inheritance was its reduction of female role to a purely passive and receptive one in the generative process.[41] But progress has been made. Stephen Pope, a theologian reflecting on the sociobiological concern to locate social and cultural behavior in the dynamics of biological emergence, asserts that our kinship with the animals actually helps humans to establish the order of personal relationships among themselves, even on the level of the theological virtue of charity. Among his conclusions are the following:[42]

First, sociobiology helps us to view human love within a much larger temporal and spatial context. The integrity of our relationships is to a large degree founded in our evolutionary heritage. Love is not a purely preternatural disposition. It is something bred into us through the dynamics of evolutionary survival and development. Such an understanding helps theologians both to diagnose the roots of antisocial behavior, and to uphold an integrity of nature that might be 'healed, perfected and elevated' by grace.

Secondly, by owning our place in an evolutionary biological emergence, we are less inclined to think of the human self as a free-floating consciousness. We are embodied in a biological dynamic and inheritance. Our loves and relationality have a biologically based emotional constitution. They are shaped in a particular direction by the genesis of nature. There are 'givens' in the human constitution; and as such, they precede freedom, and can never be repudiated, unless it be at the cost of denaturing ourselves in a fundamental manner. For instance, we do not live in an asexual or unisexual biological world—a point which critical feminism must continue to explore. Neither culture, nor spirituality, is all.

Thirdly, primary relationships to family, friends, community, society, need to be recognized in their particularity as priorities in our concerns. We belong to the whole human family through a particular family. We belong to the global community by forming a loyalty to a special place and time. Transcendence is possible only by way of the given limits.

Fourthly, our sociobiological context helps us not only to understand, but to justify priorities of concerns. What St Thomas discerned as an *ordo caritatis*, an ordering of love, founded in our natural bonds and relationships, is, to a great extent, ratified by this kind of evolutionary science.

Fifthly, the fuller flowering of altruism in a social sense, the love of our neighbor in an evangelical sense, is founded in the kin-preference derived from our animal nature. Sociobiology shows how the most all-embracing love is the confirmation of our biological nature, rather than its contrary. It is not a foreign imposition. Grace presupposes nature, in the best Thomistic style: biology, through our common animal descent and genetic inheritance, has bred into us innate affective and other-regarding orientations.

Sixthly, such a common genesis provides a basis for not only an 'emotional realization of the unity of mankind' (Scheler), but also for a deep sympathy, in the strongest sense of the word, with the whole community of living things on this planet. We feel for the other first of all not as a human rival or a potential client, nor as an animal to be tamed and used for human use, but as a companion in the unique community of life on this planet.

All this is to argue that 'We are incurably members of one another'[43]. How deeply this is so is further evoked in the following imaginative reflection:

> A good case can be made for our non-existence as entities. We are not made up, as we always supposed, as successively enriched packets of our own parts. We are shared, rented, occupied. At the interior of our cells, driving them, providing the oxidative energy that sends us out for the improvement of each shining day, are mitochondria, and in a strict sense, they are not ours. They turn out to be little separate creatures, the colonial posterity of migrant prokaryocytes, probably primitive bacteria that swam into ancestral precursors of our eukaryotic cells and stayed there. Ever since they have maintained themselves and their ways, replicating in their own fashion, privately with their own DNA and RNA quite different from ours... without them we could not move a muscle, drum a finger, think a thought.
>
> Mitochondria are stable and responsible lodgers, and I choose to trust them. But what of the other little animals, similarly established in my cells, sorting and balancing me, clustering me together? My centrioles, basal bodies, and probably a good many other more obscure tiny beings are at work inside my cells, each with its own special genome, are as foreign and as essential as aphids in anthills. My cells are no longer the pure line entities I was raised with; they are ecosystems more complex than Jamaica Bay.
>
> I like to think they work in my interest, that each breath they draw for me, but perhaps it is they who walk through the local park in the early morning, sensing my senses, listening to my music, thinking my thoughts.[44]

v. The vital level

There is no dance or art, creativity in thought or religion unless there is a basic health and wholeness in the human body and mind. To opt for high cultural values, even religious ones, or to commit oneself to some form of ecological care without a proportionate commitment to the human values of health and well-being is a monstrous self-mutilation. The dramatic situation of world poverty and the incidence of disease, e.g., the HIV virus, calls forth an enormous effort of care for the undernourished on the one hand, and for the sexual nature of our existence on the other. There is not much of a prospect for creativity in history if the rising generations are hungry, ill, and a prey to sexually transmitted disease. Physical, sexual, nutritional and hygienic education are a fundamental concern without which the human project and its healthy metabolism with the earth would be a disaster.

Hence, physical, chemical and biological values must enter into a renewed religious vision. As such values motivate a deeper and more rounded ecological concern, they also provoke a more intense wonder regarding the Providence working in the whole process which, seemingly against all odds, has brought us and this world into being. The 'higher purpose' of older considerations of the world is now illuminated as immanent in the unimaginably intricate and interconnected dynamics structuring the cosmic gestation of life itself.

The following astutely formulated principle can serve as a bridge between what we have said so far, and the observations that will follow:

> An environmental setting developed over millions of years must be considered to have some merit. Anything so complicated as a planet inhabited by more than a million and a half species of plants and animals, all of them living together in more or less balanced equilibrium in which they continuously use and reuse the same molecules of soil and air, cannot be improved by aimless and uninformed tinkering. All changes in a complex mechanism involve some risk, and should be undertaken only after careful study of the facts available. Changes should be made on a small scale first so as to provide a test before they are widely applied. When information is incomplete, changes should stay close to the natural processes which have in their favor the indisputable evidence of having supported life for a very long time.[45]

And closer to the point, to quote Schumacher:

> There are, of course, factors of production, that is to say, means-to-ends, but this is their secondary, not their primary nature. Before everything else, they are ends-in-themselves; they are meta-economic, and it is therefore rationally justifiable to say, as a statement of fact, that they are in a certain sense sacred.[46]

vi. The social level

The social level of the human good consists in the division of labor designed to produce a system promoting human welfare. The goal of such a system is an orderly recurrence of the goods and services that nourish the vital values of the community.

This is, first of all, an economic consideration: access to raw materials, the subsequent production of necessary goods, their distribution and marketing. The social value of the economic system has today become the urgent question: is it geared to the vital values of human existence, or has it lost its way in the production of superfluities or even harmful commodities? Is economics so intent on measuring everything in terms of quantity and commodity that it systematically blocks out what really counts in human well-being? The Church's 'option for the poor', the compassion and social criticism that have found expression in Liberation Theology, all forms of advocacy on behalf of the economically enslaved, the efforts of Western democracies to provide a safety net for the unemployed and the disadvantaged, evidence in their different ways a search for a new economic order where first things are kept first.[47]

The economic consideration, in turn, raises the technological issue. Is the technological apparatus of our society functioning to produce fundamental human goods, or has it become a trap, a prison, tending to make a healthy social system impossible? The compulsive production of automobiles, refrigerators, weapons, non-degradable plastics, not to mention the global proportions of the garbage dump necessary for the obsolescent, the disposable and the poisonous in our productions, has reached an impasse limiting the survival and flourishing of life.

The malfunctioning of the economic system, and the alienating influence of the technological apparatus, raise, in the most urgent manner, the question of the role of a political system in the well-being of the society. In its political determinations, society ideally expresses its concern to regulate and modify the economic system for the good of all. The political good is realized to the degree it enables the participation of the whole community in the organization of a life favorable to human culture. Its institutional arrangements are meant to ensure the rule of law and justice, thus to lift the members of society out of purely economic interdependencies; and to correct the dominative bias that thus creeps into the social structure. But when politics becomes a public relations exercise of the economic status quo, or a promotions agency for the latest technological know-how, its primary goal is forgotten. Such ever-threatening possibilities lead to a remark on cultural values.

vii. Cultural values[48]

The first thing to notice here is that cultural values are placed above political values. Politics does not tell us what we are to become; we determine ourselves. Modern feminist, pacifist, social and ecological critiques of politics keep making such a point. Politics is not the exchange of self-determination for economic well-being; nor for some form of law-and-order security. In situations of national crisis, we may well consent to certain limitations of freedom. But such a temporary compromise is made only for the sake of conserving a society in crisis for the sake of an eventual re-establishment of self-determination on the part of individuals and groups. The openness of culture, its resistance to political or economic or bureaucratic totalitarianism, is ensured only by the promotion of personal and religious values. A word on each of these.

Personal value is embodied in the self-transcending individual. To put it simply, unless there are good people doing the work of the world, with courage enough to face evil, distortion and injustice with their inner resources, no system, no politics, no government will be worth much. Such institutions will merely serve the interests of the self-seeking. On the other hand, every occurrence of honesty, integrity and moral responsibility is a seed sown for the renewal of the general culture. A recurring pattern of integrity in those who are prepared to confront the real situation, in those prepared to ask the uncomfortable questions and face uncomfortable facts, in those ready to put themselves on the line in pursuing the unpopular courses of action that the long-term common interest demands, in those prepared to opt for the common good over private advantage, secures a forward movement in the culture. For human integrity incarnates the values which animate the progress of a culture as an ongoing, life-promoting movement. I am not suggesting that this is regularly the task of the isolated individual. Social conditioning is strong, and 'structures of evil' diminish the capacities of any freedom. The individual needs others with whom to communicate in solidarity, if self-transcendence is to be a possibility. But personal value, embodied in the self-transcendence of individuals and groups, is the soul of true historical progress. From what source is that soul to draw its energies of renewal? This question leads to the appreciation of religious value.

At the heart of personal value is religious value. The sustained self-transcendence of any individual or group is mightily supported if, in the depths of human ambiguities, hope in a transcendent mystery is possible. It works in those who gently possess a peace the world cannot give. Their self-possesion is imbued with a joy that no one can take from them, just as it is inspired by a love that nothing in all creation can diminish or subvert. Religious value consists in surrendering to the grace and demands of a transcendent love. While religious value gives an intimacy with the ultimate, it breathes a spirit of self-transcending energy into human affairs. It is a love that is 'patient, kind, not envious or boastful or arrogant or rude; a love that neither insists on its own way nor is irritable or resentful'. Instead of giving way to cynical disillusionment and resting in bitter recrimination, it rejoices in the truth, as it 'bears all things, believes all things, hopes all things, endures all things. Love never ends...' (1 Co 13:4–8). It inspires a freedom that refuses enslavement to the demonic forces of the situation. It does not leave believers cowering in fear, but knows the spirit of intimacy with the healing, liberating mystery at work in the world (Rm 8:15).

Religious value is, then, a creative and a redemptive force. Its creativity is realized as a capacity to fulfill the trajectory of self-transcendence inscribed into every level of our complex being. It is redemptive because it offers the energy to begin where we are, to turn evil into good, to hope against hope, to reintegrate the fragmented elements of our existence into a new creation in remaking the conditions that favor life.

9. Summary and conclusion

These interlinking levels of the human good indicate the various ways we need to value our humanity and the world of its emergence. They pose, of course, a more profound and complex version of the question with which we began. On the other hand, the recognition of such levels of investigation provides the terms in which the issues of language and culture we raised might be subjected to much greater scrutiny. Further, the discussion of 'anthropocentricity', of the classical definition of the human, of a new scientific sense of the human and so forth, can be broken down to any number of practical questions. Above all, such distinct areas of consideration illustrate the way in which the upward and downward movement present in the connected levels of our existence are interrelated and blended. The different levels of the human good expose us to the reality of how the human is both embodied in a cosmic and planetary process while being, at the same time, an ecstatic self-transcendence homing to the ultimate creative mystery of it all.

And so, to conclude. If any of us undertook to write an autobiography, we might give expression to the movement of our lives as a pilgrim path to God, a way of many detours and false starts. Yet, withal, such a way maintains its momentum and draws its sustenance from our relationships with others. But the narrative of the somebody that each of us is would be gracelessly distorted if it did not recognize itself in the larger body of the generations: each of us is part of a generation, which in turn looks back to countless others, 'great-grandparents', our forebears... Even the most researched 'family tree' names only a tiny number of the men and women who have brought us forth. The love and courage, the struggle to survive and provide, the conflicts and the reconciliations, the language, the slow learning, the awakenings of faith and hope—all this flows into us.

Yet to halt the narrative there would be to leave out a myriad of fellow travellers, usually unnoticed, but humbly, simply, there, going along with us. True, it is not uncommon to give a little bow to these other elements as we record the love we have for our native place, for home-cooking or our favorite wine, the sunshine and the beaches, the night sky, places of vacation and refreshment, for our hobbies, books, art, even our pets, for the health we enjoyed and the fresh air we breathed. But each one of these humble elements in our story is an immense and intricate story in itself: the story of the cosmos and the earth is bred into our bones, assimilated in the food we have eaten and inhaled in the air we breathe—and received in the numberless connections that root us into the larger life of the planet.

More intimately, our material world has been formed into the world of symbols through which we read and write and speak and act, to express and explore realities in new forms and in a larger scope. A comprehensive autobiography would be one of thankfulness to the scripts already written into our brains and our bodies, into our land, into our earth, into the stars, into the whole radiating cosmos. Human beings are now in a position to tell the whole story of their origins.

We exist by grace of the past. The whole has begun to speak to us from *within* our newly elucidated experience of what has been going on. We are being taken to a fresh frontier of wonder: most of what we are, above all the *why* of what we are, the sheer fact that *we* are, can never be fully articulated. We emerge out of an ocean of sheer 'iffiness', 'chanciness', 'contingency', in ways that the ancient philosophers could never have imagined. It is, indeed, a moment in which to adore the Providence immanent in it all.

If such a past flows into us, what will emerge from us? Cosmic consciousness turns into historical conscience, into the uneasy conscience of the present generation. The alienation of man and woman, our lack of welcome for the child, our economic marginalizing of youth, the joyless irritability of our culture, the bleakness of our expectations, are all potent factors in the present. What are we handing on? What do we bequeath to the future? It is possible that we will fall toward a malice that even the Lord could not envisage (Lk 11:11), by giving our children a serpent instead of a fish, a scorpion instead of an egg; and more generally, bequeathing to the generations to come the stone of an arid and exhausted earth, instead of the bread of a nurturing and hospitable homeland. What is the prayer of our last breath as death comes to meet us out of the nature we have so long demeaned?

Human history cannot cut itself off from nature, nor is it natural for 'nature' to swallow up human history in meaninglessness, to wipe the screen clean of human meaning in blank extinction. One cannot easily believe that evolution aims at decapitation. Yet neither does it aim at disembodiment. The historical challenge at the moment is to achieve a new embodiment, a new earthing in the planetary and cosmic process. Our reflections on the past are a form of having time for the wholeness of the history and the world that have brought us forth.

If the human is, in fact, evolution become conscious of itself, then it is in the nature of the human to display its creativity by nurturing the ecological, 'the meaning of the dwelling'. If all is to be well with us, it will be because of our efforts to dwell in the larger meaning of life-giving mystery—and to really mean the well-being of our planetary dwelling.

The words which Arnold Toynbee uses to conclude his large book *Mankind and Mother Earth* provide a fitting conclusion to this chapter:

> Man is a psychosomatic inhabitant of the biosphere that coats the surface of the planet Earth, and in this respect he is one among the species of living creatures which are the children of Mother Earth. But Man is also a spirit, and, as such, he is in communication with—and in the mystic's experience, is identical with—a spiritual reality which is not of this World.
>
> As a spirit, Man possessed consciousness, he distinguishes between good and evil, and in his acts he makes choices. In the ethical field, in which Man's choices are either for evil or for good, his choices produce a moral credit-and-debit account... For the network of relations between incarnate human beings that constitutes human society, the account is still open, and will remain open so long as mankind allows the biosphere to remain inhabitable.
>
> Will mankind murder Mother Earth or will he redeem her? ... This is the enigmatic question which now confronts Man.[49]

1. For a classic treatment on the theme of the questioning character of human existence, see Eric Voegelin, *Order and History IV: The Ecumenic Age*, Louisiana State University Press, Baton Rouge, 1974, especially pp. 316–30.
2. Quoted by Gabriel Daly in his stimulating *Creation and Redemption*, Gill and Macmillan, Dublin, 1988, p. 116.
3. See Les Murray, *Blocks and Tackles: Articles and Essays 1982 to 1990*. Angus and

Robertson, Sydney, 1990, pp. 65f.

4. See George Steiner, *Real Presences*, The University of Chicago Press, 1989. His argument proposes 'that any coherent understanding of what language is and how it performs, that any coherent account of the capacity of human speech to communicate meaning and feeling is, in the final analysis, underwritten by the assumption of God's presence' (p. 3).
5. For an insightful treatment of this subject, see Neil Ormerod, *Grace and Disgrace: A Theology of Self-Esteem, Society and History*, E. J. Dwyer, Newtown, 1992.
6. Sebastian Moore, *The Fire and the Rose are One*, London, Darton, Longman and Todd, 1980, p. 67.
7. Murray Bookchin, *The Philosophy of Social Ecology: Essays in Dialectical Naturalism*, Black Rose Books, Montreal 1990, p. 116.
8. Ibid., p. 158.
9. Ibid., pp. 157f.
10. Ibid., p. 118.
11. Here, once more, I refer to James A. Nash, *Loving Nature: Ecological Integrity and Christian Responsibility*, Abingdon Press, Nashville, 1991, pp. 68–92.
12. See, for example, Thomas Aquinas, *Summa Theologica*, I, 76, 3.
13. *Commentarium in I Cor 15*, lect. 2. Brian Davies, *The Thought of Thomas Aquinas*, Clarendon Press, Oxford, 1992, pp. 205–17, gives a very clear presentation of Aquinas's hylomorphism.
14. For an abundance of wry, witty and deeply philosophical observations, see Walker Percy, *Lost in the Cosmos: The Last Self-Help Book*, Arena, London, 1983.
15. Heinz Pagels, *Dreams of Reason*, Bantam Books, New York, 1989, pp. 12f; 329. For the larger context, see Christopher F. Mooney, 'Theology and Science: A New Commitment to Dialogue', *Theological Studies* 52/2, June 1991, especially pp. 410–18.
16. Langdon Gilkey, 'Science, Reality and the Sacred', *Zygon* 24, 1989, p. 294.
17. Holmes Rolston, *Science and Religion: A Critical Survey*, Random House, New York, 1987, p. 66; also Ian G. Barbour, *Religion in an Age of Science*, pp. 146–8.
18. Interestingly, Rupert Sheldrake, *The Rebirth of Nature*, pp. 83–88; 108–10, can be read as a modern re-expression of Aristotelian hylomorphism.
19. See Jean Mouroux, in *The Meaning of Man*, trans. A. Downes, Image Books, New York, 1961, p. 256, n. 33: he is citing J. Delacroix, *Psychologie de l'Art*.
20. George Steiner, *Real Presences*, p. 19.
21. Ibid., p. 19.
22. Pierre Teilhard de Chardin, *Science and Christ*, trans. Rene Hague, Collins, London, 1965, p. 13.
23. For a fuller elaboration of self-transcendence, see Bernard Lonergan, *Method in Theology*, pp. 3–120; and for its cosmic and historical context, see Eric Voegelin, *Order and History IV: The Ecumenic Age*, especially pp. 300–37.
24. Teilhard de Chardin, *The Phenomenon of Man*, Fontana, London 1966, p. 247.
25. See Brendan Lovett, *Life Before Death*, Claretian Publications, Quezon City, 1986, pp. 77–97. For further illustration of this point, see Charles Birch, *On Purpose*, NSW University Press, Kensington, 1990. For a brief overview of the many scientific methods converging in the exploration of the 'self-organizing universe', see Arthur Fabel, 'The Dynamics of the Self-Organizing Universe', in *Cross Currents*.
26. Arthur Fabel, 'The Dynamics of the Self-Organizing Universe', *in Cross Currents* XXXVII/ 2–3, Summer-Fall, 1987, pp. 195–202.
27. Danah Zohar, *The Quantum Self*, Bloomsbury, London, 1990.
28. For an expert presentation of the new context of discussion, see John Honner, 'Not Meddling with Divinity: Theological Worldviews and Contemporary Physics' in *Pacifica* 1/3, 1988, pp. 251–72; 'The New Ontology: Incarnation, Eucharist, Resurrection and Physics', *Pacifica* 4/1, Feb. 1991, pp. 15–50.
29. See Holmes Rolston *Science and Religion: A Critical Survey*, Random House, New York, 1987, pp. 22–26 and passim. For more on the 'hierarchy of levels', Ian G. Barbour, *Religion in an Age of Science*, pp. 165–76.

30. See Arthur Peacocke, *God and the New Biology*, pp. 120–7.
31. See Rupert Sheldrake, *The Rebirth of Nature*, pp. 100–08, for another, but associated vocabulary: *holons* as successively 'nested hierarchies'. At each level, the holons are wholes containing parts which are themselves wholes containing lower-level parts, and so on. For more on 'multilevel theories', especially in reference to the work of Roger Sperry, see Ian G. Barbour, *Religion in an Age of Science*, pp. 198f.
32. Christopher Mooney, S.J., 'Theology and Science: A New Commitment to Dialogue', *Theological Studies* 52/2, June 1992, p. 319.
33. I am inspired here by Brendan Lovett, *Life Before Death: Inculturating Hope*, Claretian Publications, Quezon City, 1986. He is building on Bernard Lonergan, *Method in Theology*, pp. 47–52. For further illustration, see James G. Miller, *Living Systems*, McGraw Hill, New York, 1978; and Robert M. Doran, *Theology and the Dialectics of History*, University of Toronto Press, Toronto, 1990. This latter work is one of enormous ambition; it is particularly valuable on the profound cultural shifts that are occurring.
34. Brian Swimme, *The Universe is a Green Dragon: A Cosmic Creation Story*, Bear and Co., Santa Fe, 1984, p. 31.
35. Thomas Berry, *The Dream of the Earth*, Sierra Book Club, San Francisco, 1989, p. xiii.
36. Brendan Lovett, *On Earth as in Heaven*, Claretian Publications, Quezon City, 1988, pp. 23f.
37. A large problem is brewing for us beefeaters in the more privileged parts of the world. There are 1.28 billion head of cattle on the planet. We note that only 11% of feed goes to produce the beef itself. Cattle in feedlots produce, by consuming 790 kgs. of plant protein, only 50 kilograms of meat protein. This enormous consumption of food needed to produce so little in return, not to mention the clearing of rain forests for grazing land, and the degradation of existing cattle country, poses problems of seismic proportions. For an incisive treatment of this issue, Jeremy Rifkin, *Beyond Beef: The Rise and Fall of the Cattle Industry*. Dutton, New York 1991.
38. Mary Midgley, *Beast and Man: The Biological Roots of Human Nature*, Cornell University, Ithaca, 1978, p. 71.
39. Ibid., p. xiii.
40. The classical account is to be found in Thomas Aquinas, *Summa Theologica*, II-II, q. 26.
41. For a balanced, but perhaps surprising view of the theology of Aquinas on this point, see Catherine Capelle, *Thomas d'Aquin feministe?*, Vrin, Paris, 1982.
42. See Stephen J. Pope, 'The Order of Love and Recent Catholic Ethics: A Constructive Proposal', in *Theological Studies* 52/2, June 1991, pp. 255–88. For the conclusions, see pp. 285–8.
43. Mary Midgley, *Animals and Why They Matter*, Penguin, New York, 1983, p. 21.
44. Lewis Thomas, *The Lives of a Cell: Notes of a Biology Watcher*, Viking Press, New York, 1974, pp. 3–4.
45. Ralph and Mildred Buschbaum cited in Juan Luis Segundo in *An Evolutionary Approach to Jesus of Nazareth*, Orbis, Maryknoll, 1988, p. 39.
46. Again as cited in J. L. Segundo, *An Evolutionary Approach to Jesus of Nazareth*, p. 39.
47. For a challenging analysis, see Herman E. Daly and John B. Cobb, Jnr., *For the Common Good: Re-directing the Economy toward Community, the Environment and a Sustainable Future*, Beacon Press, Boston, 1989.
48. For a profound treatment of the following issues, see Robert Doran, *Theology and the Dialectics of History*, pp. 473–99; 527–58.
49. Arnold Toynbee, *Mankind and Mother Earth: A Narrative History of the World*, Oxford University Press, New York, 1976, p. 595f.

SECTION VI

A Sixth Circle of Connections: The Trinity

God is Life and Love as a communion of the three divine persons, Father, Son and Spirit. Any effort to relate Christian faith to a contemporary worldview would be very limited if it left out the central doctrine of the Trinity.

In the vision of Christian hope, the goal of creation is considered as being drawn into such ultimate Love-Life. In the incarnation of the Word, through the outpouring of the Spirit, the Mystery is present as gathering creation into itself. In the Trinity, the universe comes home; and all the struggling emergence of time finds its absolute future.

Before trying to tease out some of the more relevant aspects of trinitarian faith, one must register a regret that doctrinal controversies have often reduced this most comprehensive of Christian mysteries practically to the point of absurdity—at least in the minds of most Christians. What should have been a celebration of God as the absolute Being-in-Love at the heart of the universe, what might have been a sense of the divine community enfolding all conscious creation into its own love-life, appeared as an exercise in supercelestial mathematics in which one could not be properly multiplied, or divided, by three.

In recent years, no doubt out of a presentiment of new relevance, trinitarian theology has been undergoing a considerable renewal.[1] Here I shall attempt no overall statement, but merely emphasize some of the perspectives I find increasingly relevant to current ecological and cosmological discussion.

1. Ultimate reality as relational

First of all, what kind of reality is Christian faith trying to objectify out of its experience of God? How does such experience and meaning affect all our experience of reality?

Most radically, the analogical, all-connecting imagination of the Christian faith is envisaging the ultimate ground of our existence as intrinsically relational. Doctrinal theology speaks of the divine persons as 'subsisting relationships'. The divine three can only be understood in relation to one another, as 'for' and 'in' the other. The absolute one-ness of God is concretely realized in a limitlessly self-communicating relationality. God is God by being a communion of mutual self-giving. The Be-ing of God is a life of communion. And the life of God is one of unrestricted, all-embracing love.

The originating Love that God is (Father) expresses its fullness in the Word, and rejoices in its infinite excess in the Spirit. In that self-utterance and self-gift, all God is, all that the universe is or will be, is contained.

The universe emerging in the long ages of time is ever coming into being out of such Love. The eternal Now of Trinity is the matrix of time, not its contradiction.[2] That relational vitality, which theology calls the 'divine processions' of the Word and the Spirit, is creating the universe in its dynamic image. What is procession 'within' the divine mystery, is imaged in the created process of the universe.[3] The universe finds its ultimate coherence in as much as trinitarian reality draws it to participate in its own field of relationships—alight in the Word, enlivened in the Spirit, and surrendering in thankfulness to the originating and final Love. Out of its experience of this relational field present in the world through the Incarnate Word, the given Spirit and the intimation of the Father, Christian faith comes to confess the Trinity as the transcendent presence immanent in all existence. All instances of being, becoming and life have their beginning, form and goal in the 'Love-Life' of God, 'so that God might be all in all' (1 Co 15:28). The evocative language of Nicholas of Cusa returns such intense theological expressions to the world of mystical imagination: 'The Godhead is the enfolding and unfolding of everything that is. The Godhead is in all things in such a way that all things are in God.'[4]

The Word was in the beginning as the primordial self-expression of Love. God is self-differentiated into the Other, and Love becomes self-communication. Now, the universe has been uttered into existence to be a world of endlessly differentiated 'words', *logoi*, meanings. For, as Aquinas reminded us, 'created things cannot attain to the perfect image of God in a single form'. It was fitting 'that there be a multiplicity and variety in created things so that God's image be found in them perfectly in accord with their mode of being'.[5]

But the trinitarian 'self-constitution' is achieved in the Spirit. The Love that has differentiated itself, and been self-expressed in the Other, becomes now a communal activity, 'in the unity of the Holy Spirit'. Hence, we can understand the relational dynamics of creation as a participation in the ecstasy of the Spirit, leading the differentiated, distinct, and independent realities of creation into self-transcending communion.

In this perspective, the cosmos lives and breathes the mystery of 'primordial, expressive and unitive' Being-in-Love at the heart of its existence.[6] If that is true, it is very true. If the original and ultimate reality is inherently relational, if ultimate unity is self-giving communion, trinitarian faith is a healthy disturbance for all our closed little worlds of isolated independence. Defensive alienation from the other, resistance to peace and reconciliation, any hardened disharmony with the rest of creation sets us outside the stream of life.

In an obvious sense the ecological imagination is more hospitable to a trinitarian conception of ultimate reality than the former dominant mechanistic worldview. A mechanistically modelled science had little patience with any theology, let alone any theology needlessly complicated with trinitarian references to processions, relationships, the unity of the divine nature and the plurality of the divine persons. But current views of the processive and relational character of cosmos might be expected to find some of the tradition of trinitarian theology quite intriguing. The contemporary paradigms of science and the theological paradigm of God seem to be converging. As a more holistic science realizes that for centuries theology has regarded the ultimate as a realm of processive, interpersonal, relational life, with the divine reality concretely one in a manifold of relationships, points of dialogue can emerge, in the one exploration of the real.[7]

2. Productive models from the past

Generally, traditional trinitarian theology, following Aquinas, passes from the consideration of the processions and relationships of the divine persons *ad intra* (in the eternal 'within' of God) to their presence and relationships *ad extra* (in the universe of creation). The self-communication of the Trinity *ad extra* is treated under the heading of the biblical category of 'mission' or 'sending'—the way, for instance, God 'sends' the Son and Spirit into the world. While current theological efforts tend to find such categories too abstract and spatial[8], I am beginning to suspect that there is something to be gained, given the present evolutionary horizon, in attempting to retrieve some aspects of it. In giving a brief indication of this point, I will be taking some liberty with the traditional terminology, but only in the hope of suggesting a better comprehension of its meaning in the present context.

The following points might be made. For example, Aquinas asks how a divine person can be 'sent'. The problem here is to imply neither inequality in the coequal Trinity, nor some primitive form of spatial movement—as though God were not everywhere in the first place. Here, Aquinas isolates the two relevant points: first, the divine person is sent inasmuch as the eternal procession of the Son/Spirit is prolonged into time and history. The life, the consciousness of God thus takes in the world of time and its emerging world. Secondly, because the mission is an extension into time of the eternal procession, it means that the divine persons begin to exist in the world in a new way: 'Thus, the Son is said to have been sent by the Father into the world, even though he was already in the world, because he began to be in the world in a visible way by taking flesh.'[9] Through the divine missions the Trinity immerses itself in the created universe. It is experiencing itself, as it were, in all the risk, fragility, and process of a temporal world.

Thus the divine missions are the Trinity's dynamic openness to the world and to history. The Mystery communicates itself to the created other in order to enfold creation into its communal life.

But there is a second point. For Aquinas allows for two dimensions of these missions, the 'invisible' and the 'visible'. Now, I feel that failure to transpose this traditional distinction in an adequate manner has locked Christianity into a narrowness that is ill-prepared for the cosmic scope of its present challenge.

So, first, a word on 'the invisible missions'. We begin with the recognition that God is present to everything and everyone in an absolutely fundamental manner, as the sheer Be-ing, *Ipsum Esse*, the source of all being. As the giver of existence, God is thus present in the innermost depths of all reality—as we described in a previous chapter. But, out of the depths of the ocean of Being in which we exist, there comes a wave of freedom, of self-communicating love. God is not content, as it were, simply to be the nameless universal mystery at the heart of reality. Beyond giving ourselves to ourselves, the Trinity wishes to give itself to us; thus to be the realm of a new selfhood and communion. This is the area of grace, of a new gift, in which we are not only God's creatures but God's intimates. God acts in creation out of the fullness of the self-communicating mystery, to be present to creation in the nakedness and special connectedness of trinitarian Love.

Now this implies the transformation of human consciousness traditionally termed 'sanctifying grace'. Human consciousness is enkindled in a special experiential immediacy with the divine.[10] It is 'conformed to God' and configured to the divine persons in the love of the Spirit and in the wisdom of the Word: 'because by such knowing and loving, created consciousness (*rationalis creatura*) makes contact with God in its activity; in this special manner God is said not only to be in such a consciousness, but to dwell in it as in his temple' (1, 43, 3). Thomas will also speak of the Father as given and indwelling through grace, but not as 'sent'. He is present as the ultimate source and goal of this self-communication, 'above all and through all and in all' (Ep 4:6).

Thus, 'invisibly', as an horizon of loving presence preceding any particular context of time or space, God indwells human consciousness. However unnamed, unexpressed, in an 'invisible' depth of life, human consciousness is awakened into a new level of being, to participate in the very consciousness of God as self-communicating love. Our capacities for dialogue are now extended to loving communication with the divine: 'for the Word is not any kind of word, but the Word breathing love.'[11] The implication is that wherever there is evidence of a consciousness lovingly alive, in reconciling wisdom and self-transcending love, it is the dwelling of God.

True, this metaphor of 'sending' might all too easily give the impression of the divine persons arriving from the outer space of the divine realm. On the other hand, once the spatial limits of the metaphor are recognized, the reality is more like a blooming or an emergence of the mystery out of the depths of existence. What could have been, or appeared to be, a universe of simple fact, however uncanny, is now irradiated as a field of divine consciousness: God dwells in us and we in God in the circulation of divine life and love.

But there is more. The divine missions are not only 'invisible' as the Word and Spirit illuminate and enkindle the indefinable horizon of life. They are also 'visible'. The Divine Word becomes incarnate; while the Spirit is manifested as a movement in history, in transformed lives, in communities of faith, in all the signs and sacraments of Church. The reality of God's self-giving attunes itself to the embodied reality of our mode of being. It takes shape in the time and space of this particular world. The divine mystery does not inhabit a transcendent celestial realm apart from the flesh-and-blood of our world; nor, says Aquinas, do we:

> Now it is connatural for a human being to be guided by the visible to the invisible. Through the evidence of creatures, God has in some way manifested himself and the eternal processions of the persons. Likewise, it is right that the invisible missions of the divine persons be made manifest through visible creation. Now this occurs in one way with the Son, and in another way with the Holy Spirit: ...the Son is visibly sent as the author of holiness; but the Spirit as the witness to that holiness.[12]

For Aquinas, the embodied world of our existence is the natural milieu for human communication. So, it is within this given ecology, one might say, of human existence, that trinitarian love manifests itself to communicate with who, and what, we are in the concreteness of our coexistence. This implies that the created world, in all its differentiation, structures, relationships, and dynamics, already evidences its primal emergence from the Be-ing and relational reality of God (the processions): a trinitarian dynamism is originally inscribed into every element of the created world. It is fitting, then, that the character of its trinitarian orientation be made manifest within it. God thus respects the ecology of human consciousness. Christ enacts and expresses, in the world of human communication, the meaning of the mystery at work. And the Spirit—symbolized in the water, the wind, the air, the flame of our natural experience—is expressed as the field of new connections.

In this perspective, the Church is an extension of the visible missions, as that part or moment of the world that is expressly alive to the universal mystery of relational love at work. The community of Christian faith deals in words and signs and symbols and sacraments of the trinitarian mystery. It is part of the ecology of God's presence, not the totality of it.

Secondly, that 'connatural' world of human communication that Thomas recognized as the place of God's express self-manifestation is now recognized as a time-structured evolutionary process. The primordial dynamism of the processions, their extension into creation through the invisible and visible missions, now demand, if God's communication is to be revealed in its full connaturality, a re-expression in an evolutionary perspective.

In such a perspective, the invisible missions are 'invisible' precisely because they energize the indefinable totality of the universe. Animated by such energies, the emergent universe can be contemplated in the light of one vast act of divine self-communication. The self-giving God becomes the soul, the heart, the mind of the world's emergence: 'There is... one God and Father of all, who is above all and through all and in all' (Ep 4:6). On the other hand, the 'visibility' of the missions suggests a sense of the universe in terms of God's embodiment. The mystery of Love incarnates the Word in the whole cosmic body of creation as the form and structure of creation. It en-worlds itself as Spirit in all the self-transcending dynamics that characterize our cosmic becoming. And, in the presence of 'the Father', intimated as the source of both Word and Spirit, it offers itself as the absolute future of the whole of temporal existence.

In this way, the trinitarian God is 'enworlded' as the relational ground of a cosmos of growing connections and relationships. If the Word is incarnate as universal meaning, if it is first a question, its full expression is in a desire welling up from the depths of the incarnation: 'As you, Father, are in me, and I in you, may they also be in us... may they be one as we are one' (Jn 17: 21–23). And in a promise: 'When the Spirit of truth comes, he will guide you into all truth... and he will declare to you the things that are to come' (Jn 16:13).

Realism, of course, demands an acknowledgement of the tragic world of conflicts in which we live. Whatever our eschatological hopes, this is no whole world of harmonious connections. For that reason, we must bear in mind that visibility of Love in such a world must first of all be healing and reconciling, before it is simply transforming. Divine self-giving is marked with the intensity of self-sacrifice. The Godself communicated in the Son and the Spirit is directed to human beings locked in an apparently insuperable problem of evil. Incarnation goes to the point of crucifixion. And the Father is revealed as having no self-disclosure in this world other than the Crucified One. Similarly the Spirit is exposed to the world of self-enclosure, of non-relationality, by witnessing to no power and no truth other than that of the selfless love of the Cross. God's visibility in the world, then, connects with our problem of evil. It is love revealed as exposed to, but ever greater than, the alienation, failure and fragmentation of our world. If it promises transformation, it is first of all a healing for the self-destructiveness of our ways.[13]

Though such a suggestion is cryptic in the extreme, it does at least suggest the possibilities of reworking the traditional schema of processions, divine relations, invisible and visible missions in a very fruitful manner.

3. An extended frame of reference

There is nothing new in exploring the reality of God by way of analogy with created realities, nor in interpreting the world of created reality in the light of God. Augustine's 'vestige' and image doctrine, with its underpinning in Greek exemplarism, is a good instance.[14] In Catholic tradition, it developed into the typically Franciscan cosmic sense of all reality sacramentally manifesting God. For example, over seven hundred years ago, Robert Grosseteste of Paris could contemplate a speck of dust to find in its existence, form and goodness a manifestation of the three divine persons.[15] Now, a speck of dust caught in the light beaming in through a medieval scholar's window is rather different from other specks we know today: say, the earth as a speck in a universe of some hundred billion galaxies, or that speck of incredible compression from which the universe has been unfolding for fifteen billion years since the Big Bang. Still, theology can find new ways to contemplate the originality, the dynamic form, the wonder and beauty of such realities; and, in so doing, come to a fuller insight into the deeper implications of its central mystery.[16]

Likewise, St Bonaventure, in his ladder of contemplation, could consider creation as implying a trinitarian presence on three ascending rungs of intensity. First, the divine footprints (*vestigia*) in the prepersonal; secondly, the divine image in the spiritual; thirdly, the divine likeness brought about in the transformation of grace:

> The creation of the world is a book in which the creative Trinity shines forth. It is read according to the three levels of expression, the ways of vestige, image and likeness. By these, as up a ladder's steps, the human intellect has the power to climb by stages to the supreme principle which is God.[17]

The implied metaphor of a stable ladder leading upward tends now to be replaced by the arrow of time moving forward in an evolutionary direction. The 'footsteps' of God now have to be tracked in the history of the fifteen billion years of the world's emergence. The 'image' of God is now disclosed in terms of human consciousness emerging in such an evolutionary world as a light and a love in which the universe is known and appreciated. The 'likeness' of God, in its turn, is disclosed in the transformation of that consciousness as it indwells and is inhabited by the primordial mystery of creative Love.

For his part, Aquinas sums up the themes of the great Greek theologians, especially Athanasius and Gregory of Nyssa, when he treats of the presence of the Trinity in the act of creation:

> ... the divine persons are causes of the creation of things in the order of their procession, since God acts from his knowledge and will as a craftsman acts in regard to what he produces. The craftsman acts through a word conceived in his mind and through love in his will in reference to what is to be made. Hence the Father creates through his Word which is the Son, and through his love which is the Spirit...[18]

In such a vision, the trinitarian mystery is involved in the very mystery of what it means to be a creature. The trinitarian processions of Word and Spirit are at the foundations of the cosmic process. Indeed, Aquinas goes on to say that:

> The knowledge of the divine persons is necessary for us for two reasons: in the first place, to have a right sense of the creation of things. Because we say that God created all things by his Word, the error is excluded of holding that God produced things out of the necessity of his nature. Because we hold that there is a procession of love within him, it is clear that God did not produce creatures out of some extrinsic need, but on account of love for his goodness... In the second place, and this is more important, that we might have a proper sense of the salvation of the human race, which is brought about by the incarnation of the Son and through the gift of the Holy Spirit.[19]

That 'right sense of creation' implies that creation is not a divine necessity, but an act of love, a communication from the heart of God. Secondly, the 'salvation of the human race' is based on the self-communication of God to the cosmic, evolutionary process, just as it is energized into a final unity 'through the gift of Spirit', the Holy Breath and atmosphere of 'life to the full'.

In such a perspective, trinitarian theology exploits all the resources of language to explicate the trinitarian form of creation. Because it cannot rest content with an undifferentiated sense of the presence of the mystery, it explores the variety of ways in which its presence can be discerned in the relational and processive meaning. Hence, trinitarian theology has traditionally employed an evocative technique known as 'appropriation'. General considerations of the divine attributes are given a trinitarian focus; typical examples, in reference to the Father, Son and Spirit respectively, are: power, wisdom and goodness; unity, equality and harmony; eternity, beauty and form; omnipotence, omniscience and will; efficient, exemplary and final causality, and so on. With such a plethora of terms, trinitarian thinking tries to give some expression of the encompassing mystery of God 'ever beyond us', as Father, all-inclusive Source and Goal; of God now 'with us', as Word and Son incarnate; of God ever 'within and between us' as Holy Spirit, pervading all creation.

Such expressions anticipate a new set of trinitarian connections drawn from the book of creation as read through modern eyes. Many writers speak of the direction of evolution in terms of increasing 'differentiation, subjectivity and communion'.[20] The initial cosmic event unfolds into a marvelous variety of particles, forces, elements, life-forms, cultures. With that differentiation, there occurs an increase of interiority and consciousness. Living, self-organizing unities emerge into growing complexity, through the nervous system to the human brain. In the human phenomenon, consciousness emerges as intelligence and freedom. Evolution becomes conscious of itself. The universe is self-aware. And in that self-awareness, the mystery of our common origins and shared destiny comes to dwell: communion.

Trinitarian theology could make many connections with these three principles. For example, the mystery of the self-differentiation of God through Word and Spirit could be understood to be the primordial implication of the principle of cosmic differentiation. The manifold of creation emerges out of God's original self-utterance and joy in being.

Likewise, the development of interiority is climaxed in being drawn into the divine realm of consciousness. In faith, hope and love, human consciousness participates in the trinitarian vitality of self-presence: we begin to know and love ourselves and the universe of divine creation as God does. Interiority thus culminates in dwelling in the divine mystery, and its indwelling of our minds and hearts.

Finally, creation is not only grounded in the self-differentiation of God; not only does it participate in the interiority of the divine self-knowledge and love, but it expands in the field of divine communion. The self-giving relationality of God's Being-in-Love affects all created differentiation and subjectivity with its own unity-in-difference. To be, to live, to develop has the trinitarian meaning of consciously being, living and developing from, and for, and with the other. Trinitarian communion is, thus, the limitless field in which interpersonal, ecological and cosmic communion can be realized. In short, our planetary existence is a living image of the Trinity, and a progressive participation in the communal life of the divine mystery.[21]

A spiritual writer provides us with a remark that brings many of these points together. She is in a long tradition when she says:

> The cosmos has all the marks of the Trinity: it is a unity; it is internally differentiated but interpenetrating; and it is dynamic, giving, expansive, radiant. And, as a work of art, the cosmos has another important character: it does not exist for the sake of something else, something beyond itself; it is not useful, it is not instrumental; it is an end in itself, self-justifying, valuable in its own right and in its very process. This, I think, is foundational for...ecological virtue...[22]

4. Conclusion

A vivid sense of the trinitarian dimension of human existence profoundly affects the lived sense of human selfhood. Christian theology speaks of the divine indwelling, at once God dwelling in us, and ourselves dwelling in God. To search into who we are is to find ourselves in the presence of God, the Self in all our selves. The three classic biblical expressions of such intimacy with the divine mystery are that we are temples of the Holy Spirit, the divine creativity hidden in all creation; we are members of the Body of Christ in whom all things cohere; we share in the divine life as sons and daughters of the Father as the final all-welcoming mystery of the future.

Such an indwelling means that God is known with an 'inside' knowledge; a knowledge of participation in the Love-Life that God is: 'Beloved, let us love one another, because love is from God; everyone who loves is born of God and knows God... For God is love... No one has ever seen God; if we love one another, God lives in us and his love is perfected in us' (1 Jn 4:7–12). To experience God in such a way is to find oneself as a 'connected self', a self-to-be-realized in relationship to the other. This 'other' today admits of a global, ecological and cosmic extension. Hence the adoration of such a God orients the believer into a world of relationship and communion. It implies an agenda for the transformation of ourselves, our communities, our global coexistence.

If, in this brief reflection, we have moved trinitarian faith from the background to the foreground of our thinking, it is not for the sake of needless complexity. But in a world of both needless complexity as well as astonishing connectedness, Christian faith can find a new health and a new wholeness in contemplating the universe in the light of its fundamental mystery.

1. For a considerable bibliography and a fuller expression of my own approach, see my *Trinity of Love: A Theology of the Christian God,* Michael Glazier, Wilmington, Delaware, 1989.
2. For a rigorous examination of the theme of divine eternity, see John C. Yates, *The Timelessness of God*, University Press of America, Lanham, Maryland, 1990.
3. I do not think this distinction is sufficiently appreciated. When it is not, 'process' becomes the ultimate reality, not 'procession'. There is a big difference between deifying the process and adoring the 'proceeding' divine persons. Even a survey of such towering expertise as Ian G. Barbour, *Religion in an Age of Science*, op.cit., does not mention the relevance of the trinitarian mystery to current evolutionary and genetic understandings of reality. Remaining only with the category of process strikes me as being theologically far too timid. In contrast, see the strong trinitarian emphasis of the scientist Bernhard Philberth, *Der Dreieine*, Christiana-Verlag, Stein am Rhein, 1974.
4. Quoted by Rupert Sheldrake, *The Rebirth of Nature: The Greening of Science and God*, op.cit., p. 198. Noteworthy, too, is the influence of Sheldrake on the deep trinitarian structure of Bede Griffith, *A New Vision of Reality,* op.cit.
5. Thomas Aquinas, *Summa contra Gentiles*, book 2, ch. 45.
6. Here I have adapted three terms taken from the great work, John Macquarrie, *Principles of Christian Theology*, SCM Press, London, 1977, pp. 195–202.
7. For example, see the various references to the Trinity in Fritjof Capra, David Steindl-Rast and Thomas Matus, *Belonging to the Universe*, pp. 61–63; 106–09; 118; 131f.
8. Including my own, as in *Trinity of Love*. See pp. 133–72 for evidence of a different method.
9. Thomas Aquinas, *Summa Theologica*, 1, 43, 1.
10. Ibid. 1, 43, 5 ad 2.
11. Ibid. 1, 43, 5 ad 2.
12. Ibid. 1, 43, 1. For further treatment of the missions, see Jean-Hervé Nicolas, O.P., *Synthèse dogmatique: de la Trinité à la Trinité*, Beauchesne, Paris, 1985, pp. 231–65.
13. See my *Trinity of Love*, pp. 168f; 195–202.
14. St Augustine, *De Trinitate* VI; VIII–XV.
15. Servus Gieben, 'Traces of God in Nature according to Robert Grosseteste', in *Franciscan Studies* XXIV, 1964, pp. 154f.
16. For the profound analogical shift in our understanding of God's relationship to the world, a most useful article is Carol Jean Vale, S.S.J., 'Teilhard de Chardin: Ontogenesis vs. Ontology', *Theological Studies* 53/2, June 1992, pp. 313–37. The author implies that a more critical comprehension of the Thomistic tradition of analogy would make for a greater continuity between Aquinas and Teilhard.
17. St Bonaventure, *Intinerarium Mentis in Deum*, c.3, n.5.
18. Thomas Aquinas, *Summa Theologica* 1, 45, 6 ad 2.
19. Ibid., 1, 32, 1 ad 3.

20. For example, Thomas Berry, *The Dream of the Earth*, Sierra Book Club, San Francisco, 1988, pp. 44–6.
21. See my *Trinity of Love: A Theology of the Christian God*, pp. 8–13. For related biological applications, see Rupert Sheldrake, *The Rebirth of Nature*, pp. 193–9. Also, Fritjof Capra, reflecting on the 'self-organizing' properties of matter manifested in 'the pattern of organization, the structure, and the process', remarks, 'The funny thing about the concept of self-organization is that it can be presented as having a "trinitarian" nature.' (*Belonging to the Universe*, p. 117). Bonaventure and Thomas Aquinas would not have found it so odd; indeed, they would have been delighted, I am sure, by such a remark.
22. Beatrice Bruteau, 'Eucharistic Ecology and Ecological Spirituality', in *Cross Currents* Winter 1990, p. 504.

SECTION VII

A Seventh Circle of Connections: The Eucharistic Universe

The eucharist as the source and goal of the whole life of the Church is necessarily the most compact expression of Christian coexistence and relationality. The trinitarian field of relationships is embodied in sacramental symbolism. For the eucharist celebrates and enacts a reality that is at once cosmic in its extent and quite earthy in its symbols.[1]

1. Nature, culture and eucharistic reality

Hence, the mystery unfolds: the real presence of our ultimate being 'in Christ' comes to us through the shared 'fruit of the earth and the work of human hands'. In this light, the eucharist intensely symbolizes a holy and whole creation: nature and history, the produce of the earth and the productions of human creativity, the self-expression of the human and the self-communication of God are subsumed into the one sacramental intensity.[2]

In a world threatened with ecological disintegration and cultural uprootedness from nature, the eucharist nourishes faith into a gracious, ultimate wholeness. It is a paradigmatic instance of the Pauline kerygma, 'God was in Christ reconciling the world to himself' (2 Co 5:19). In the 'holy communion' offered in the eucharist, not only does God communicate himself in the body and blood of Christ, but also, in a profound sense, gives us back to ourselves. Our humanity is nourished into an ultimate awareness of its embodiment in a material cosmos.

The most intense moment of communion with the divine is at the same time the most intense moment of our communion with the earth. The fruits of the earth and the works of human hands are not magically vaporized by the action of the Spirit: they come into their own as bearers of the ultimate human mystery. Put most simply, the bread and wine become 'real food and real drink' to use the

Johannine idiom. 'Transubstantiated' in this way, the sacramental elements anticipate the cosmic transformation that is afoot, not as something that leaves the created cosmos behind, but as promising its healing and transformation.

This is one of the important ingredients that Christian hope can offer to ecological consciousness. It seems we need more than a purely rationalist ethic that is hard put to prescribe more than a conservation of resources. A naive ecological ideology tends to regress to a past and unattainable innocence; while a doctrinaire evolutionism empties the significance of the present into an impersonal and incalculable future. For its part, Christian hope, nourished on its sacraments, envisions an absolute future actually occurring within our earthly and historical present.

The great emancipations of the modern age have had to pay a particularly heavy price. In the struggle against what was perceived as oppressive tradition or archaic order or biological limitation, such forms of liberation have also left so much behind. The modern emancipated individual is uprooted from any sense of nature, and detached from any sense of a sacred nurturing universe. The Enlightenment had its costs, and to the degree Christians were willing to pay up, the range of universal connectedness was lost in the exchange. Mircea Eliade remarks:

> As for the Christianity of the industrial societies and especially the Christianity of intellectuals, it has long since lost the cosmic values it still possessed in the Middle Ages. The cosmic liturgy, the mystery of nature's participation in the christological drama, have become inaccessible to Christians living in a modern city... at most we recognize that we are responsible to God and also to history—but the world is no longer felt as the work of God.[3]

The alienation of no longer being able to feel that the world is the work of God is extreme. The resources of Christian vision and sensibility exist somewhere to be retrieved into a more enlightened Enlightenment, this time more attuned to the nature and cosmos of our existence. A cosmic connection was, indeed, part of the deep sensibility of a Ptolemaic vision of universe, as, for example, in the points of medieval theology we have mentioned. But now, after the Newtonian mechanist objectification of the cosmos, human consciousness is in need of a reimmersion, as it were, in the cosmic whole. This does not mean a return to the Ptolemaic universe. It is more a matter of recovering the meaning, both humble and hopeful, of the ancient declaration of Ash Wednesday, ' Remember, man, that thou art but dust and unto dust thou shalt return'. But now that character of that dust must be understood as a cosmic reality. A vast change in sensibilities is implied as we come to understand ourselves as made of stardust: 'each of us is a distillation, a condensed centrifuge of cosmic energy'.[4] We are embodied in a cosmic totality.

2. Universal transubstantiation

Admittedly, when we begin to reflect on the eucharist in this larger context, we find that a good deal of eucharistic theology tended to congeal in categories pertaining to what has often been termed a 'two-tiered' version of reality.[5] A crude, excessively polemical notion of transubstantiation left little room for the cosmic imagination. The bread and wine were simply 'the matter' of the sacrament. The 'substance' of such realities is simply replaced by the reality of Christ. The cosmically unnerving aspect of that kind of transubstantiation is that the 'real presence' of Christ ousted the 'real presence'—the substance—of the bread and wine. The heavenly Christ, coming from what was crudely imagined as the 'outside' of the world, replaces the essential form of what was on the inside, the substance of the bread and wine. Only the 'accidents' remained to figure as a provisional sign of another reality. The upshot was that, though the sacrament was held 'to contain' the heavenly Christ in a mysterious fashion, the earthly realities concerned no longer had lasting significance.

While not denying the intense realistic thrust of such a version of eucharistic faith, I believe it can be expressed in a far more ample manner, more apt to do justice to the cosmic mode of Christ's presence in the one universe of reality.

We need a larger cosmic perspective. The limitations of analogical thinking are always there; but still, it is truer to the mystery of the eucharist to set it within a cosmic process of 'transubstantiation'.[6] The physical, the chemical, the biological, lead, through their successive transformations, to human consciousness—the universe aware of itself as living mystery, as awaiting, as tending towards its ultimate connection: 'We are nature's big chance to become spirited'.[7]

For Christian faith, the range of such 'spirited-ness' is finding its ultimate extension. The movement of progressive transubstantiation peaks within human history. In the Spirit-inspired 'Let it be with me according to your word' (Lk 1:38)[8], Mary embodies the genetic potential of creation, to become pregnant with its final life-form: 'Let the earth be open to bud forth the saviour!' In bringing forth the Christ, into life on this planet, into life unto the end in the Cross, into life transformed in the Resurrection, creation is subsumed into a field of trinitarian life: 'May they be one, as we are one'(Jn 17:22). 'Spiritedness' opens to an horizon of the self-communicating of Love. 'Nature's big chance to become spirited' is enfolded in the Spirit's action in forming the Christ.

As the Spirit leads human consciousness to live in the light of the Life Incarnate, Christian faith blossoms into its sacramental imagination: symbols, gestures, words, relationships and biological processes of our world are appreciated in different contexts as 'visible signs of invisible grace' (Augustine). These reach their most intense and comprehensive form in the eucharist. The risen Lord takes the broken pieces of creation, elements of our earthly reality, which nature and history have combined to produce by a long process of transubstantiation, to turn them into something more: the anticipation of a new totality: 'This is my body, this is my blood...'.

In one of the great theological works of the modern era, we find a paragraph to focus the meaning of what we have been trying to say so far:

> The mythical understanding of the world sees the whole world as a sacred theophany. In an eschatological sense, this is also what the world is for Christian faith. If the cosmos as a whole has been created in the image of God that appears—in the First-Born of creation, through him and for him—and if this First-Born indwells the world as its Head through the Church, then, in the last analysis, the world is a 'body' of God, on the basis of the principle not of pantheistic but hypostatic union.[9]

It is such a transforming identification that faith must continue to enact: 'Do this in memory of me'. The Lord invites us to reconnect with the cosmos as he has done, to claim it as our own, as our larger selves, in a world of divine incarnation.

3. Eucharistic contemplation

The early differentiation of God and nature that Hebrew faith and Greek thought achieved is now being led back, not as a regression, but as a new stage of mature identification with the all, the whole. Connections with the path of mystical prayer are suggested.[10] Contemplation begins with a disciplined withdrawal from creation for the sake of the Absolute. Detachment is preliminary to openness and surrender. The way of negation, John of the Cross's repeated *nada*, 'nothing, nothing, nothing... and even on the mountain nothing'[11], leads to emptiness and silence. Beyond all conception, image and projection, we are left only with what is there, what is given, with only...God—and what can God do. To pray is to be thus 'nakedly intent' on the divine will, and lost in it alone, beyond the tiny clamor of needs and manipulation: 'it is to your advantage that I go away...'(Jn 16:7).

But then the illumination. In going up the mountain of ascent, we meet the Other coming down into the plains of creation: ever creating the world, offering it to us, incarnating himself within it; above all, inviting us to join in what God is doing in our world. To Philip's request, 'Lord, show us the Father and we will be satisfied', the Word Incarnate answers, 'Have I been with you all this time, Philip, and you still do not know me? Whoever has seen me, has seen the Father' (Jn 14:8f). But even here the play of transcendence and immanence continues: the clearing of a space for the transcendent mystery prepares for a new experience of the immanence of God: '...if I do not go away, the Advocate will not come to you; but if I go, I will send him to you' (Jn 16:8).

In such a dialectic, we begin to know our place for the first time. The real presence is disclosed as really real—in the concrete earthly reality of our being in the world, a point beautifully made by a man of faith who is also an outstanding poet, as he reflects on Jesus 'praying in a certain place' (Lk 11:1):

> When Jesus prayed, and taught us to pray, he was doing two things. He was entrusting himself to the Begetter of the universe, and he was giving himself to his brothers and sisters throughout time and space. What we call the 'Our Father' or the 'Lord's Prayer' faces at once into all that has ever surrounded and determined the fortunes of the human race, and into the lives of individual men and women. It frames an act of confidence in the goodness of God who made and makes us, it articulates our shared need, and it declares a resolve to be creative on behalf of others. The name of the 'certain place' of prayer is, in other words, 'Compassion'. In that country of the mind and heart, one sees that our universe is not an anonymous indifferent milieu, but the homeland and heartland of God. And at the same time, one sees that God the life-cherisher calls all of us to be life-cherishers and life-givers in concert with him. We ask for food and forgiveness, because we need them both. We agree to offer food and forgiveness, because others need them both. If we mean what we say, prayer will send us back, a little shaken but more than a little heartened, to the tasks of everyday... The country of Compassion can be, and should be, wherever we happen to be.[12]

An overwhelmingly intricate and vast, impersonal objectification of the universe has been achieved in recent centuries of scientific exploration. It has been attained as an intellectual conversion—away from the myths, the symbols, the common imagination of prescientific stages of human culture. The objectivity of science, intent on presenting minds with the universe as it is, broke clean away from human subjectivity.

Now that objectivity needs itself to be converted. The impersonal whole must be reattached to the subjectivity from which it arose. The world of objects must be reclaimed as a universe of relationships, if we are not to define ourselves out of it 'all', and allow scientific exploration to empty the cosmos of its most obvious and astonishing phenomenon, human consciousness itself. An objective world ignorant of, or ignoring, consciousness is a world without meaning, without value; fundamentally without persons. It is a world of fragments and dislocation, rather than a universe of communion and connectedness.

Precisely at this point, in an horizon determined by the incarnation of the divine in our midst, the eucharist draws us back to a personal universe of relationship and communion. The objective 'all', symbolized in the earthly, material elements of the shared bread and wine of the sacrament, is celebrated as in a state of transformation into a new universal reality. The Body of Christ becomes the milieu of our existence. Through the eucharist, the whole is now offered to be reclaimed as *our own*. In eucharistic imagination, the whole reality of the world is imagined 'otherwise'. It redeems the superficial, the fragmented, the alienating elements in our experience into another vision: a Christened universe, seeded with the Spirit-energies of faith, hope and love, being transformed—transubstantiated—into the Body of the Risen One.[13]

4. The Real Presence in the real world

Thus, eucharistic faith envisages existence in the world as an indwelling in shared mystery. It invites us into a transformed field of communication. Though we human beings have been busy through our short history with mutual recriminations, the Divine Word has been writing our collective name in the ground of our common earth at the beginning of a second 'cosmic day'. For 'in Christ', as the Pauline vision sees it, 'all things hold together' (Col 1:17), and are gathered up in him (Ep 1:10). Though the 'image of the invisible God', he is 'the firstborn of all creation' (Col 1:15). All things are made 'in him', and in their finality, 'for him' (v.16). The mystery of Christ is of cosmic comprehension, inscribed into the origin, the end, the consistency of the universe of God's creation. Once more we can leave von Balthasar to provide a powerful summary statement:

> But in his definitive form, he [Christ] takes up into himself all the forms of creation. The form which he stamps upon the world is not tyrannical; it bestows completeness and perfection beyond anything imaginable. This holds for the forms of nature, concerning which we cannot say... that they will simply disappear, leaving a vacuum between pure matter and man, who

> is a microcosmic fruit of nature. To be sure, it is only in man that nature raises its countenance into the region of eternity; and yet the same *natura naturans* that in the end gives rise to man is also that *natura naturata*, and the whole plenitude of forms which the imagination of the divine nature has brought forth belongs analytically to the nature of man. The same holds in greater measure for the creations of man in his cultural development: they, too—they especially!—belong to him as the images he has produced out of himself... to impress upon the world and which have a continued existence in man by reason of their birth even when they have perished in time. The same, finally, holds to a supreme degree for the creations in the realm of grace.[14]

Such a sweeping and intense statement accents how all reality—the physical world, all forms of life, the distinctive life of human consciousness, its cultural creations, and its transformation in the Spirit—is *in* and *for* Christ, embodied in the plenitude of the Risen One.

Set within such a comprehensive vision, the transformation of the fruit of the earth and work of human hands into the body and blood of Christ anticipates a cosmic transformation. The new 'substance' of the transformed bread and wine anticipates the radical 'transubstantiation' of all creation, when the inner consistency and coherence of all things in Christ will be achieved.[15] The bread and the wine, the matter of the sacrament, are products of nature and culture. But in Christian imagination, in the field of Spirit, they are given back to us as the real presence of a transfigured cosmos.

Thus, the community of hope comes to communicate in the ultimate form of human and cosmic reality. The bread and wine are not abolished by the Spirit's action. Rather, they are constituted in their ultimate significance, now food and drink to nourish an ultimate form of existence. They are no longer 'the food which perishes, but... the food which endures to eternal life' (Jn 6:27). The eucharist is the 'true bread from heaven' (Jn 6:32), 'the bread of God which...gives life to the world' (v.33), 'the bread of life' (v.35): as nourishing us with the life-giving being of Christ, the food of the eucharist is 'my flesh for the life of the world' (v.51).

In short, the eucharist is the sacramental communication of what is 'food indeed', 'drink indeed' (v.55). Through the transformative action of the Spirit, the eucharistic elements are no longer mere nutrients of biological life ('not such as the fathers ate and died..' (v.58), but the food and drink of eternal life. In their meaning as real food and drink for the community of believers, the eucharistic bread and wine provide a foretaste of life in a transformed universe.[16] As Christian faith ingests such a sacramental reality, human existence is no longer understood as merely consuming the reality of this world, but as being nourished into a definitive relationship to 'all things in Christ'.

Though each sacramental celebration occurs within the particularities of space and time, it can be understood as enacting a eucharistic process of cosmic dimensions. The following oft-cited declaration by Teilhard de Chardin strikingly expresses such a vision:

> When the priest says the words *Hoc est corpus meum* ['This is my body'], his words fall directly on the bread and directly to transform it into the individual reality of Christ. But the great sacramental operation does not cease at that local and momentary event... these different acts are only the diversely central points in which the continuity of a unique act is split up and fixed, in space and time, for our experience. In fact, from the beginning of the Messianic preparation up till the Parousia, passing through the historic manifestation of Jesus and the phases of growth of his church, a single event has been developing in the world: the incarnation, realized, in each individual, through the eucharist.
>
> All the communions of a lifetime are one communion. All the communions of all human beings now living are one communion. All the communions of all people, present, past and future, are one communion...
>
> In a secondary and generalized sense, but in a true sense, the sacramental species are formed by the totality of the world, and the duration of the creation is the time needed for its consecration. *In Christo vivimus, movemur et sumus* ['In Christ, we live and move and have our being'].[17]

5. A eucharistic ecology

How do we begin to tease out the ecological ramifications of such a vision? The evolutionary and ecological emergence of our being discloses indeed that we are 'up to our necks in debt' to life.[18] How do we begin to repay what we owe in a non-inflationary currency? What are the conditions of an equitable exchange? The eucharistic command of the Lord, 'Do this in memory of me', arises from the imagination of one whose existence is unreservedly relational... They point in a certain direction.

'Do this..' at least means having a heart for it all, putting our souls back into the shared body of our coexistence; having time, beyond the instant demand, for the wholeness to emerge. It means letting what was previously disowned enter our hearts, to be claimed in the relationships of a larger self. It means welcoming the mystery of the cosmos in a more generous hospitality, and into a larger spiritual space where the all, the whole, is not abandoned to absurdity, despair or defeat. Finally, it is a matter of owning nature and all its processes as our own flesh and blood. We enter the stream of time not only as jubilantly participating in the feast, but also, through all the givings and service that life and love demand, to be part of the meal. We are destined to contribute our living energies to the great banquet of existence; to be, with him, the grain of wheat falling into the holy ground to die; and so, not to remain alone (Jn 12:24).

The planetary consequences of such a eucharistic vision have been beautifully expressed by a contemplative writer who appreciates:

> the Earth as the Eucharistic Planet, a Good Gift planet, which is structured in mutual feeding, as intimate self-sharing. It is a great Process, a circulation of living energies, in which the Real Presence of the Absolute is discerned.[19]

Again, we return to the point: from one point of view, the eucharist is faith, hope and love imagining the world 'otherwise'. From another, it is the objective symbolism of faith nourishing our minds and hearts into such an imagination, and away from the sickness of individualist, self-serving fantasy. The 'real food' here is indeed strong meat for our sickly constitutions, weakened over time as they have been by the junk food of cultural egoism. The need for a more salubrious nourishment is expressed in the words of Einstein:

> A human being is part of the whole, called by us the 'universe', a part limited in time and space. He experiences himself, his thoughts and feelings, as something separated from the rest—a kind of optical delusion of his consciousness. This delusion is a kind of prison for us, restricting us to our personal desires and to affection for a few persons nearest to us. Our task must be to free ourselves from this prison by widening our circle of compassion to embrace all living creatures and the whole of nature in its beauty.[20]

The relational existence commanded by, and embodied in, the Christ of our faith lives from a sense of reality notably at odds with any individualistic vision: 'Instead of taking as the norm of Reality those things which are *outside* one another, he takes as a standard and paradigm those who are *in* one another.'[21] This is instanced in the exalted prayer of Jesus, '... that they may all be one. As you, Father, are in me and I in you, may they also be in us... I in them and you in me, that they may be completely one' (Jn 17:21f).

Existence, by implication, is 'holy communion'. It is characterized by mutual nourishment and indwelling. We are 'in' one another for the life of the other. In that coexistential sense of being is found the true meaning of the personal, in the trinitarian sense, and in the planetary demands of the present.

These eucharistic insights converge to suggest a deep ecological sensibility. For the eucharist in its deep religious meaning situates ecological awareness in a gracious field of shared life and living relationship. The word *eucharistia* is Greek for 'thanksgiving', the first movement of Christian existence, giving thanks for the wonder of the love that has called us into this communion of life. As such it nourishes the heart and imagination into the kind of 'thanking' that will deeply condition the 'thinking' necessary to address the urgent ecological problems of our day.

How the sacramental sense of eucharist is anticipated by ecological concerns is picked up by one noted ecologist in the following words:

> To live, we must daily break the body and shed the blood of creation. When we do this knowingly, lovingly, skilfully, reverently, it is a sacrament. When we do it ignorantly, greedily, clumsily and destructively, it is desecration. In such a desecration, we condemn ourselves to spiritual, moral loneliness, and others to want.[22]

How the eucharist can nourish us into such an ecological and sacramental sense must remain a concern for the church of the future. Here I have merely given an indication of possible directions. However, one thing is clear: it is time to think of the community eucharist as a rededication not only to sharing the bread of life with the hungry, not only as a communion with the suffering, but also as a common commitment to the great sacrament of the earth itself. It is the central point from which the Church can enter with its deepest imaginings into that movement described by one of its most critical exponents as:

> ... humanity becoming more fully integrated with the being of Gaia, more fully at one with the presence of God. It is a deepening into the sacramental nature of everyday life, an awakening of consciousness that can celebrate divinity within the ordinary, and, in this celebration, bring to life a sacred civilization.[23]

6. Conclusion

I can sympathize with those who might feel that religious symbolism is one thing, while the conflicts and strategies of practical ecological concerns are quite another. I can only suggest that the movement toward a richer and more inclusive life begins with a new way of imagining the world. Great symbols orient us within the wholeness of things, and give both the passion and patience to grapple with it. Here, I have offered a reflection on the eucharist as a primary symbol within the life of Christian faith. It is an essential expression of the poetry of such faith, unfolding as it does in a universe of grace. Like all poetry, it makes us cherish, in the words of David Malouf:

> all those unique and repeatable events, the little sacraments of daily existence, movements of the heart and intimations of the close but inexpressible terror and grandeur of things, that is our *other* history, the one that goes on, in a quiet way, under the noise and chatter of events and is the major part of what happens each day in the life of the planet, and has been from the very beginning. To find words for *that*, to make glow with significance what is usually unseen, and unspoken too—that, when it occurs, is what binds us all, since it speaks immediately out of the center of each one of us; giving shape to what we too have experienced and did not till then have words for, though as soon as they are spoken we know them as our own.[24]

My modest contention here has been that the eucharist, as essential to the primary poetry and imagination of faith, illumines the 'little sacraments of daily existence', to immerse us more deeply in that 'other history' of immense unfolding. It gives shape to what we experience, and makes the unseen and unspoken glow with significance, even if the struggle to have words for such matters remains...

1. For a more general treatment, see my *Touching on the Infinite: Explorations in Christian Hope*, Collins Dove, Melbourne, 1990, pp. 129–53.
2. For a Protestant comment, see Paul Tillich, 'Nature and Sacrament', in *The Protestant Era*, Chicago Univesity Press, Chicago, 1984, pp. 94–112.
3. Micea Eliade, *The Sacred and the Profane: The Nature of Religion*, trans. Willard R. Trask, Harcourt, Brace and World, New York, 1959, p. 179.
4. See the very imaginative and inspiring reflection of David S. Toolan, 'Nature is a Heraclitean Fire', in *Studies in the Spirituality of the Jesuits* 23/5, November 1991, pp. 1–46. The phrase cited is on p. 37. Also, Denis Edwards, *Made From Stardust*, Collins Dove, Melbourne, 1992.
5. For a lot of valuable background material and for an abundance of stimulating suggestions, see John Honner, S.J., 'A New Ontology: Incarnation, Eucharist, Resurrection, and Physics', *Pacifica* 4/1, February 1991, pp. 15–50. I would agree, of course, that the 'two-tier' universe of reality is hardly commensurable with the emerging 'quantum view', but I would argue that such a dualism is very much a nominalist and modern perversion of the paradigm of analogical participation characteristic of Aquinas and the great Scholastics. Here, I think, there remains a great deal to be profitably transposed into the epistemology of quantum physics.
6. For a wide-ranging and, I think, seminal work, see Gustave Martelet, *The Risen*

Christ and the Eucharist World, trans. René Hague, Crossroad, New York, 1976.
7. D. Toolan, 'Nature is a Heraclitean Fire', p. 36.
8. Note the youthful enthusiasm of her *genoito* (the intensive optative) in contrast to the more passive form of *genestheto.*
9. Hans Urs von Balthasar, *The Glory of the Lord: A Theological Aesthetics. I: Seeing the Form*, trans. E. Leiva-Merikakis, J. Fessio and J. Riches (eds.), T. & T. Clark, Edinburgh, 1988, p. 679.
10. See Beatrice Bruteau, 'Eucharistic Ecology and Ecological Spirituality', in *Cross Currents* Winter 1990, pp. 499–514. Despite the avowedly eclectic approach of this writer, her reflection is a profound meditation on the relevance of classic Christian doctrines to ecological awareness.
11. See Crisogono de Jesus, et al (eds.) *Vida y Obras de San Juan de la Cruz*, Biblioteca de Autores, Cristianos, Madrid, 1955, p. 492.
12. Peter Steele, S.J., 'Praying in a Certain Place', unpublished, quoted with permission.
13. See D. Toolan, 'Nature is a Heraclitean Fire', pp. 36–43. Clearly, I am indebted to the formulations we find here, even if my emphasis is somewhat different.
14. Von Balthasar, *The Glory of the Lord*, pp. 679f.
15. Here I am indebted to F.X. Durrwell, *L'Eucharistie, sacrement pascal*, Cerf, Paris, 1981. In what follows, I have relied especially on pp. 77–113. Also of special value is G. Martelet, *The Risen Christ and the Eucharistic World*, trans. R. Hague, Seabury, New York, 1976, especially pp. 160–97.
16. See Geoffrey Wainwright, *Eucharist and Eschatology*, Epworth, London 1971. This book is a rich resource even though the author is not sympathetic to the line we have taken, e.g, pp. 104f.
17. Teilhard de Chardin, *The Divine Milieu*, Harper and Row, New York, 1960, pp. 123–6.
18. A happy phrase borrowed from D. Toolan, 'Nature is an Heraclitean Fire', p. 43.
19. B. Bruteau, 'Eucharistic Ecology and Ecological Spirituality', in *Cross Currents*, Winter 1990, p. 501.
20. Albert Einstein, quoted in Michael Nagler, *America Without Violence*, Island Press, Covelo, California, 1982, p. 11.
21. B. Bruteau, 'Eucharistic Ecology and Ecological Spirituality', p. 502.
22. Wendell Berry, *The Gift of Good Land*, North Point Press, San Francisco, 1981, p. 281.
23. David Spangler, *Emergence: The Rebirth of the Sacred*, Dell Publishing Co., New York, 1986, p. 81.
24. David Malouf, *The Great World*, Picador, Sydney, 1990, pp. 283ff.

SECTION VIII

Dimensions: Death and Love

A. The Way of Death

I must confess to a certain uneasiness in finding so little mention of death in recent 'holistic' writing—even though there are significant exceptions.[1] It will be well to reflect for a few pages on the universal fact of death. If a holistic view of life or an evolutionary vision of the cosmos remained silent on this aspect of our existence, we would be right to suspect the influence of some kind of reality-denying fantasy undermining an energetic commitment to life.

At any moment, it is likely that our awareness will find itself poised between two classic statements on death. One expresses an immemorial sense of tragedy: death is an inescapable and all-engulfing fact: *Sunt lacrimae rerum et mentem mortalia tangunt* [2]—roughly translated as 'there are tears at the heart of things, and all that dies deeply affects our souls.' The other, from John's Gospel, is the statement of Jesus: 'Unless the grain of wheat falls into the earth and dies, it remains just a single grain; but if it dies, it bears much fruit' (Jn 12:24). Here, too, there is an inevitable sense of suffering implied; but it is suffering taken up into an ultimate hope for transformation and communion. Let us delay for a moment in the places where these two visions play.

1. 'The tears of things'

It requires no special depth of perception to find easily in others, with difficulty in ourselves, that our most driven forms of consumerism and exploitation arise out of a secret terror. We do not want to die. Material hoarding, endless insurance, addiction to power, to sexual conquest, to work itself, the frenzied cultivation of health and beauty, are all fairly obvious strategems deployed to keep the mortal enemy at bay. All such efforts inhabit a world of fearful fantasy. The self is packaged in what we so pitifully take as a deathless image. With subterranean cunning, the psyche tries to keep a hold on life as it is, precisely because it neither finds nor trusts a movement in life to something more. The stream of life is going nowhere; there is nothing there to catch us up and bear us on.

Now, these rather common attempts to deny mortality can migrate into a new, perhaps more benign ecological or cosmological form. Nature is so beautiful, such an explosion of life and variety, that nothing ever dies, or is allowed to do so. There is no place for death in the scheme of things. Death is not natural; and hence must be banished from life and concern as the obscene intruder. The cosmos of nature is cosmeticized.

What I am trying to suggest here is not an exercise of arrogant analysis, as though any of us occupies either an immortal or unambiguously hopeful standpoint. After all, the contemporary imagination is infested with a great variety of images of extinction, and who remains unaffected?[3] It is difficult to give death its simple due when deadly things have demanded so much of our psychic energies this century. The *Shoah*, Hiroshima, the still-continuing nuclear threat, the ecological crisis in all its manifestations—the thinning ozone layer, the Greenhouse effect, air pollution, the degeneration of the cities, acid rain, the poisoning of the waters and oceans, the destruction of rain forests, the erosion of the soil—all have their effect.

Such lethal realities affect feeling, imagination, our whole taste for life. Add to them the bleak objectification of life achieved through modern technologies of production, control and surveillance, and you have the roots of a sense of life divorced from any pattern of meaning, save that of mere survival. As economic structures break down, we may well feel lost in it all, mere appendages to the machine, blips on the computer screen, engulfed by the system which calls no one by name, and cherishes nothing except itself. When we live in 'a post-catastrophic world', in turn massaged and irritated by a haze of media images of the predominantly tragic 'news', some scientific authorities do nothing to lift the heart to a more comprehensive vision:

> man is the chance genetic mutation: to accept such a message is to awaken from a millenary dream into a total solitude and isolation in the universe quite alien to human self-worth, a world that is deaf to human music, indifferent to human hopes, sufferings and crimes.[4]

Such a passage is a good expression of the prevailing cultural atheism of our day. In the experience of death, it cashes its cheques. Death is no longer experienced as passing over into God, but as meaningless extinction, the dead end. The promise of life, of love, of beauty, of intelligent exploration and moral responsibility, simply cannot be kept.

The sense of individual extinction appears almost insignificant compared to the manifold possibilities of collective death. Though human history has always known its catalogue of natural disasters, famines, earthquakes, plagues—'acts of God'—we now live with the eerie possibility of 'human acts' producing a planetary suicide. The possibilities are there: biological warfare, thermonuclear incineration, ecological collapse. The vast intricate ecosystem of nature seems to be going through a gigantic spasm, whether of death-throes or of healthy purging might be too soon to say. Science, on which Western culture has so pinned its hopes, has been revealed as not necessarily user-friendly; or, more seriously, not necessarily accompanied by values proportioned to the earth-transforming power it wields. In such cultural distress, art appears merely as a cry of pain, religion a naive projection, mysticism nothing but a regression to the innocent ignorance of the child; while scientific exploration becomes a sophisticated journey into the void.

Death has indeed become more deadly, and it is little wonder that our minds and hearts recoil in repressive dismay from its presence. Even the sexual relationship, where nature is experienced at its most ecstatic and creative, can no longer be trusted. The HIV virus has put an end to that. For its part, evolutionary optimism has to confront the laws of entropy, and the eventual heat-death of the cosmos itself, 'a scenario that many scientists find profoundly depressing'.[5]

2. The denial of death

So much of modern culture with its glorification of youth and beauty, its obsessive consumerism and its schizoid individualism, arises out of a denial of death. Our way of life is marked with the pretence that we are outside the domain of true creaturehood, with its inherent finitude and mortality. As Ernest Becker remarks in his classic reflection entitled *The Denial of Death*:

> The idea of death, the fear of it, haunts the human animal like nothing else; it is the mainspring of human activity—activity designed largely to avoid the fatality of death, to overcome it by denying in some way that it is the final destiny for man.[6]

Becker acknowledges the enormous influence of Freud in the understanding of psychopathology. But he is critical, too. Relying on the insights of Søren Kierkegaard and Otto Rank, he lays bare a certain inconclusiveness in Freud's approach when it touches on the matter of death and the deep terror that it strikes into every human being. For his part, Becker argues that the fundamental repression or denial in human life is not sex, as Freud had taught, but death. Whatever therapy might be brought to bear on human problems, it is only freeing us to live with the most radical fear of all: that which has its origin in our mortality. What are we to do with the fact that we are on our way, inescapably, to death?

Negative though such a question might sound, it poses deeper questions about the urgent ecological matters and the larger cosmic connections we have been exploring. Are these further examples of the denial of death, or do they contain within them a more healthy acceptance of our connectedness to the whole? As Becker highlights the primordial terror experienced in the human psyche in the face of death, he is inviting us to a radical acceptance of creaturehood. We are immersed in the larger wholeness of nature, connected to it, caught up and carried along by it. The paradox is that only by accepting our limitation and contingency within a larger whole, only by yielding ourselves into the life-process, can we arrive at true freedom and psychic health.

Religious experience is essentially a 'creature feeling', in the face of the massive transcendence of creation. It registers a sense of being located as a tiny, vulnerable instance within the overwhelming miracle of Being. At this juncture, religion and psychology find a talking point—'right at the point of the problem of courage.'[7] Faced with the immensity of the universe and its apparent impassivity in regard to individual fate:

> ...man had to invent and create out of himself the limitations of perception and equanimity to live on this planet. And so the core of psychodynamics, the formation of the human character, is a study in human self-limitation and in the terrifying costs of that limitation. The hostility to psychoanalysis...will always be a hostility against admitting that man lives by lying to himself about himself and about his world, and that character, to follow Ferenczi and Brown, is a vital lie.'[8]

The larger ecological and cosmic connections can either expose the 'vital lie' of culture, or, by denying death in their own way, feed it with further self-deception. What new sense of self might such connections offer? How can they be genuinely creative of a deeper, more wonderful participation in the mystery of life? How can they heighten, rather than diminish, a sense of reality and responsibility?

In contesting the evasiveness of modern consciousness in the face of death, Becker summons us to the courageous acceptance of our creatureliness as the only authentic way:

> By being or doing we fashion something, an object or ourselves, and drop it into the confusion, make an offering of it, so to speak, to the life-force.[9]

Thus, in effect, he is laying the foundation for a lived sense of relationality and for the praise and thankfulness that are at the heart of religious faith. A shared sense of the human condition consequently demands renewed collaboration between science and religion. Science improperly absorbs all truth into itself. It is the role of religion to stand for a larger version of truth, enabling human beings to:

> wait in a condition of openness toward miracle and mystery, in the lived truth of creation, which would make it easier to survive and be redeemed because men would be less driven to undo themselves and would be more like the image which pleases their creator: awe-filled creatures trying to live in harmony with the rest of creation. Today we would add ...they would be less likely to poison the rest of creation.[10]

To live with a genuinely creaturely consciousness is to relativize the repression and denial at work in human consciousness. To the degree we accept our puniness in the face of the overwhelmingness and majesty of the universe, to the degree we become aware of the unspeakable miracle of even a single living being, and waken to the chaos or 'panic' in the immense, inconclusive drama of creation, we come to a point of healing.

Following Frederick Perls, Becker detects four protective layers structuring the neurotic self. The first two layers are the mundane, everyday layers of cliche and role. Doubtless, it is some achievement to break out of image of the self communicated to us in the individualism and consumerism of the day. But:

> ...the third is the stiff one to penetrate: it is the impasse that covers our feelings of being empty and lost, the very feeling we try to banish in building up our character defences. Under this layer is the fourth and the most baffling one: the 'death' or fear-of-death layer; this...is the layer of our true and basic animal anxieties, the terror that we carry around in our secret heart. Only when we explode this fourth layer... do we get to the layer of what we might call our 'authentic self': what we really are without shame, without disguise, without defences against fear.[11]

The 'authentic self' of connectedness to the whole, of ecological care and cosmic hope, is only gained by befriending the mortal character of our existence. The authenticity consists in the healing which comes from regaining the attitude of humility—a sense of radical finiteness, our creatureliness.[12]

In short, the great value of Becker's work lies in its revaluation of humility, in the most original sense of the word. If the ecological and cosmic connections we have been exploring are to be something more than posturing, the cultivation of humility as a basic attitude to life and creation is a necessity.

As a word, humility derives from the Latin *humus*, meaning the 'earth', 'soil', 'dirt'. It points to the existential fact that our life is earthed, grounded, bound up with immense dynamism of nature into whose processes we are each and all immersed. Hence, the ancient liturgical injunction: *Memento homo quia pulvus es...* : 'Remember, man, thou art but dust...' The virtue of humility—for it is a *virtus*, in the moral sense, a quality of freedom and self-determination—enables human consciousness to deal creatively with the dread of death, by 're-membering' ourselves in the ground of the universe. Death is permitted to emerge from its subterranean place of influence, no longer able to sap our energies or drive us to the frenzy of illusory immortality-projects, but to connect us with the whole, and to immerse us in a universal good.

Out of this trembling acceptance of our mortality can come the wisdom we need. Life remains a question, within an overwhelmingly questionable universe; and because each of us is a question, what is most clear is that none of us is the center of that universe. We have emerged out of a vast cosmic process and are dying back into it. When we begin to ask about the true center, the true life-force of this overwhelming universe, an acceptance of self as mortal and finite begins. With that, if not precisely adoration, at least a surrender to the unnamed, incomprehensible Whole can be realized. Only a de-centered self, conscious of its mortal limitation, can live for a larger mystery.

3. The grain of wheat, the fertile ground

'The wisdom we need...'. The Christian taste for life is, as I mentioned above, contained in the words of Jesus: 'Unless the grain of wheat falls into the earth and dies, it remains just a single grain; but if it dies, it bears much fruit' (Jn 12:24). Surrender to, participation in a larger vitality, giving oneself into the ground of the whole mystery, transformation into an ultimate coexistence, are all implied here. To enter into the 'chaos' of dying is to rise to a new level of being. It is to be drawn into the 'white hole' of Jesus' resurrection, the whole of creation transformed by the Spirit.

The following words, taken from a notebook of Dostoievsky, are a significant expression of creation made whole on the personal level:

> To love somebody else as one loves oneself, which Christ told us to do—that is impossible. We are bound by the force of earthly personality: the 'me' stands in the way. Christ, and Christ alone, did it; but he was the eternal ideal, the ideal of the ages, to which one aspires and must aspire, impelled by nature.
>
> Nevertheless, since Christ came to earth as the human ideal in the flesh, it has become clear as daylight what the last and highest stage of the evolution of the personality must be. It is this: when our evolving is finished, at the very point where the end is reached, one will find out... with all the force of our nature that the highest use one can make of one's personality, of the full flowering of one's self, is to do away with it, to give it wholly to any and everybody, without division or reserve. And that is sovereign happiness. Thus, the law of 'me' is fused with the law of humanity; and the 'I' and 'all' (in appearance two opposite extremes) each suppressing itself for the sake of the other, reach the highest peak of their individual development, each one separately. This is exactly the paradise that Christ offers. The whole history of humanity, and of each individual man and woman, is simply an evolution towards and an aspiration to, struggle for, and achievement of, this end.[13]

The basic entropy affecting an individual biological existence is dissipated to allow for a higher realization of communion. The occlusion of the mystery of the all, which structures a death-denying life of mere survival, is transformed into the open circle of 'life to the full'.

Analogies from life and science abound, especially when a scientist combines scientific exploration with a mother's experience. In a chapter on the 'survival of the self' into which she extends her 'quantum view of the self'—male consciousness more akin to the particle model, female more like the 'wave' aspect—Danah Zohar concludes:

> My own experience of the truth of the process view came through the experience of pregnancy and early motherhood, but one needn't be a mother, or even a woman, to appreciate the essential connectedness of quantum theory and what it is telling us about ourselves as quantum systems. We all... have a feminine side, a 'wave aspect', an aspect which surrenders rather than grasps, which 'gives itself up' to things beyond the nuclear self rather than concentrating on building boundaries around the self. This is the side we must cultivate if we are to transcend isolation and the consequent, and needless, terror of death...
>
> The kind of surrender required to make the most of quantum

> process is like Christ's saying, '... and whoever will lose his life for my sake will find it' (Mt 16: 25). On a quantum view, he who would find himself a place in eternity must fully wed himself to life's processes now.[14]

The dynamism of the 'self-transcending self' explored in so many ways in what has gone before, finds itself inscribed, underwritten as it were, in the upward vector of an ascent from electron, to atoms, to molecules, to proteins, to cells, to organisms, to the complexity of the human brain, and to the cosmic overture of human consciousness. The direction of life is one of transformation. Might not death be more hopefully located in such a process, not as a dissolution, but as the expansion of the self into its fullest relationships. In that way, death would not be an alien intruder, but a relative—'Sister Death' as St Francis could pray—within the universal process.

For, in the larger scale of experience, events occur in which we are taken out of ourselves, to plunge us in the deeper stream of life. A culture affected by the 'denial of death' tends to see life more in terms of a slowly melting block of ice, followed by a final and definitive heat-death. Entropy conquers all, even the universe itself. And yet, life does go on: in the sudden turns of wonder, of surprising insight, of joy; in the occurrence of great loves and of humility before the strange grandeur of moral achievement. In such moments, there is an uncanny 'more' in the experience of the mystic, the artist, the martyr, the prophet, the great thinker, the scientist. We are in the presence of eternity in the making, of 'eternity coming to be as time's own mature fruit'.[15]

Such instances dramatize what all of us can feel, perhaps more tentatively, as life's true direction. Death is given its place, not as the destruction of the self, but as dying out of all the limitations of an ego, as yielding into the deathless dimensions of the real. The following words of wisdom catch the point:

> If you surprise the world with your life, the world will surprise you at death. Don't think of death as extinction; such uninspired speculations are simply too prosaic to be true. Your dull imagination insults the very grandeur and staggering wonder of the universe...embrace your death. It will serve you.[16]

So it is that a meditation on death is set within the larger process of meditating on life within the emergent universe. Death becomes a matter of dying into the ultimate mystery of life, as a necessary transformation in the unfolding of the cosmic mystery.

Perhaps that sounds too much like simple optimism—ultimately repressive of the true deadliness of death. A complex problem remains. How can hope find a psychological focus that is neither depressive nor schizophrenic? If it is morbid to constrict the whole hopeful direction of life to a meditation on death, it would be just as evasive to repress the piercing tragedy at the heart of our existence. The mystery of death demands that we hold together both the negative and the positive dimensions of our experience. Yet these extremes don't simply meet; for there is no all-knowing comprehensive standpoint, no deathless theoretical vision. How then, can they be best held in hopeful realistic tension?

In that transformation of consciousness we call Christian hope, we find a symbol which can allow both meanings to exist, and to illumine each another: the crucifixion and death of Christ himself.[17] The meditation of the Christian is not fixated on a skull, but on the cross of Jesus. In it deepest meaning, it is a theophany: the all-creative mystery reveals itself as compassionate love. In the deadliness of Jesus' death—as failure, isolation, condemnation, torture—transcendent love has become familiar with our problem of evil.

But not to be defeated by its power. For the death of the crucified embodies the ultimate form of life as self-surrender to its all-inclusive mystery. Jesus' unreserved dedication to the realm of true life, 'the Reign of God' and his solidarity with the defeated and the lost, are concentrated in a final point of self-offering. It is precisely at that point that God is self-revealed as a love stronger than death, as the creative mystery that holds in being and fulfills all the best energies of life. Thus, the transformation of the Risen One, the 'white hole' in the world of death.

True, this is no resuscitation to our present biological life, no relocation in the time and space of this world, no cure for death. Only a transformation of human existence can answer the hopes written into a life: the entropy, the limiting individuality of biological life, is definitively overcome: a new creation is enacted in Christ as 'the way, the truth, the life'. In the four Gospel narratives, the empty tomb is the historical marker of the cosmic transformation that has begun in Christ. That emptiness and darkness highlight the ultimacy of what is now embodied in Christ:

> So if anyone is in Christ, there is a new creation: everything old has passed away; see, everything has become new! (2 Co 5:17).

Hope, nonetheless remains hope. It is never immune to the inconclusiveness of experience. It lives always in the 'in-between' of what is, and what is yet to be. It must wait on the mystery of complete transformation. For even the New Testament writer soberly concedes, 'As it is, we do not yet see everything in subjection to him' (Hb 2:8f).

Yet for all the sobriety of Christian hope, its great fact remains. In

Christ, the universe has been changed. It has been radically 'Christened'. As, too, the reality of death. Christ did not die out of the world, but into it, to become its innermost coherence and dynamism. In his death and resurrection, the mystery of the incarnation is complete. For the Christian, to yield into the reality of such a death, to surrender to the transforming power of such a resurrection, is to be newly embodied in the future form of cosmos. The struggle and dying we now know is moving in that direction:[18]

> The last enemy to be destroyed is death...When all things are subjected to him, the Son himself will also be subjected to him who put all things under him, that God may be all in all. (I Co 15:26–28)

The mortality of our existence leaves us poised over an abyss of life. The empty tomb, with its implication of the transformation of physical matter into the Spirit's creation, has a cosmic significance.[19] As a sign of the full-bodied reality of resurrection, the emptiness of the tomb seeds our history with questions and wonder: what great transformation is afoot? And also with doubt—in those who resign themselves to a closed-system or vicious circle view of the world. The empty tomb, so soberly recorded in each of the four Gospels, offers no salvation in mere emptiness. It functions as a factor within the awakening of faith. For that new consciousness unfolds, first as caught up in the movement from the tomb (discovered as a puzzling fact) to the cosmic surprise of what had happened (in the appearances of the Lord), then back to the tomb (as an index of a new creation)—then out into the limitless horizons of a transformation of all things in Christ. Such faith is not primarily looking back at a death, but facing forward into the promise of life, in a universe transformed.

A fitting epilogue is provided by Karl Rahner, as we catch him pondering the cosmic significance of the Christian mystery:

> Christ is already at the heart and center of all the poor things of this earth, which we cannot do without because the earth is our mother. He is present in the blind hope of all creatures who, without knowing it, are striving to participate in the glorification of his body. He is present in the history of the earth whose blind course he steers with unearthly accuracy through all victories and all defeats onwards to the day predestined for it, to the day on which his glory will break out of its depths to transform all things. He is present in all the tears and in every death as the hidden joy and the life which conquers by seeming to die... He is there as the innermost essence of all things, and the most secret law of a movement which still triumphs and imposes its author-

ity even though every kind of order seems to be breaking up. He is there as the light of day and the air are with us, which we do not notice; as the secret law of movement which we do not comprehend because that part of the movement which we ourselves experience is too brief for us to infer from it the pattern of the movement as a whole. But he is there, the heart of this earthly movement and the secret seal of its eternal validity. He is risen.[20]

1. Albert LaChance, *Greenspirit: Twelve Steps in Ecological Spirituality*, Element, Rockport, Massachusetts, 1991, pp. 99–124.
2. Virgil, *Aeneid*, I, 462.
3. See the profound analysis of Robert Jay Lifton, *The Future of Immortality*, Basic Books, New York, 1987.
4 Jacques Monod, *Chance and Necessity*, Knopf, New York, pp. 172f.
5. Paul Davies, *God and the New Physics*, Penguin, London, 1983, p. 204.
6. Ernest Becker, *The Denial of Death*, The Free Press, New York, 1973, p. ix.
7. Ibid., p. 50.
8. Ibid., p. 51.
9. Ibid., p. 285.
10. Ibid., p. 282.
11. Ibid., p. 57.
12. Ibid., p. 58. See Andras Angyal, *Neurosis and Treatment: a Holistic Theory*, Wiley, New York, p. 260.
13. Quoted by Yves Congar, *The Wide World My Parish*, Darton, Longman and Todd, London, 1961, pp. 60f.
14. Danah Zohar, *The Quantum Self*, Bloomsbury, London, 1990, pp. 134f.
15. See Peter C. Phan, *Eternity in Time: A Study of Karl Rahner's Eschatology*, Associated University Press, London, 1988, p. 55. Also, pp. 53–8; 207–10.
16. Brian Swimme, *The Universe is a Green Dragon: A Cosmic Creation Story*, Bear and Co., Santa Fe, 1984, p. 117.
17. Gustave Martelet, *L'au-delà retrouvé. Christologie des fins dernières*, Desclée, Paris, 1975, pp. 33–98.
18. For elaboration of this point, see Denis Edwards, *Jesus and the Cosmos*, Paulist Press, Mahwah, New Jersey, 1991, pp. 103–32. The author here presents and extends the insights of Rahner. See, too, for more on Rahner's theory, Peter Phan, *Eternity in Time*, pp. 79–134; and on a more popular level, Marie Murphy, *New Images of the Last Things*, Paulist Press, New York 1988, pp. 13–19.
19. See John Polkinghorne, *Science and Creation: The Search for Understanding*, SPCK, London, 1988, pp. 64–68. See also, my *Touching on the Infinite: Explorations in Christian Hope*, Collins Dove, Melbourne, 1991, pp. 112–16; and for a rich theological context, J. Moltmann, *The Way of Jesus Christ*, SCM, London, 1990, pp. 250–62.
20. Karl Rahner, 'Hidden Victory', in *Theological Investigations VII*, trans. David Bourke, Darton, Longman and Todd, London, 1971, pp. 157f.

B. Wholesome Sex

If I began reflecting on death by noting how seldom it is being treated in recent literature on the great connections we have been considering, something similar is the case in regard to sex. One must wonder at the fragility of any comprehensive vision of the 'All and the Whole' when two such elemental facts of life are so little addressed. The least we can suspect is that a potentially new sense of human existence is running the risk of being stunted, rendered unearthly, by such neglect. A death-less, sex-less vision of the cosmic emergence would be simply unreal.

1. A conflict

Sex has been around a long time; and in this post-Freudian unrepressed era, it is certainly a highly differentiated, even dominant, feature of Western culture. But there the oddness of the situation begins, and the reality becomes a vast and perplexing question. Why has sex been so unrelated to the universal picture?

That is a strange question. For, in the unitive and generative capacities of sex, reality is experienced as ecstatic, life-affirming, relational. Sexual union involves the couple in the stream of the generations. It places them within the universal and planetary generative process. It is an act of communion both affirming, and affirmed by, a larger cosmic embodiment. The creative range of spirit is earthed in the relational structure of the sexual body. The point of the aphorism is inescapable: 'the body is spirit *incognito*' (Sandor McNab).[1]

Of course, the poetry and love-songs of the world know all about that. Nature smiles on lovers. Religion blesses marriage. And a baby's smile disarms the cold heart. Everyone feels better. Love, life and the beauty of the world are of a wonderful piece.

And yet, the problems. For those prepared to think about the wider implications of our sexual existence, the cosmic and ecological dimensions of sexuality are swallowed up in the urgent issues of overpopulation and the practical management of the power of sexuality in ourselves and others—especially the young. Here, disease, unwanted pregnancies, sexual exploitation, economic distress, troubled marriages, and family breakdown dominate the discussion.

One aspect of the problem is that this seemingly most intimate and ecstatic activity of our embodied nature has often become grotesquely objectified. Experts discourse on techniques; advertising's lurid images displace the elementally earthy reality; the porno movie is the turn-on; the dominant concern is the performance, enacted to the criteria of a voyeuristic culture. How does such artificiality allow intimacy with either person or nature to survive?

Further, the structure of the modern economy does not easily allow for the intimacy of the generations, of the family, of the couple. The motivations of individual achievement and success dominate the soft concerns of 'private life'. Relationships to place, to people, to the intimate other must yield to economic demands. Here, permanence of commitment is so often harried by mobility of career. The impersonal marketplace provides the measure of success: the 'rule of the household' (economy) does not take in the generative values of the household itself. When a weak family structure is the condition imposed by the all-demanding economic scene, the family tree has, necessarily, very disturbed roots. The intimate identities of the home are replaced by the roles and masks of economic performance.

Because of the unrelatedness of modern technological life to nature as a whole, it is little wonder that sexual activity emerges as 'the only green thing left' in the experience of the technopolis. Cut off from larger bondings, uprooted from the process of nature, the range of intimacy tends to be compacted into sexual activity. Culture begins to suffer an erotomaniac fixation. Where past times struggled, often harshly, for some form of sublimation of the sexual drive, our own time has been uniquely liberated for reduction of everything to the sexual. In a de-natured and de-sacralized world, there is precious little of the sublime to recommend any sublimation.

The currently regulated 'sex industry' is a social manifestation of such fixation. Its wider legitimation is promoted by much of modern advertising and 'adult entertainment'. If you do not accept your place in a society of voyeurs, it is likely that you will feel subject to a cultural sexual harassment of a most pervasive and intrusive kind! But, of course, conscience stirs; or, at least, nostalgia for a more wholesome past: child abuse, child pornography, paedophilia are beyond the pale; and various other forms of sexual aggression and exploitation are the subjects of insistent lamentation—even if, within the limits of a permissive culture, it is difficult to formulate why the line should be drawn at any point.

When so much of experience is so intensely concentrated in the sexual, the larger problems of our relationship to nature, hitherto deferred or ignored, begin to come home in a sexual guise, to shatter the innocence of sexual liberation. Tragic limits make their presence felt as in virulence of the AIDS epidemic. Sex becomes dangerous, and whole departments of public health are taken up with stratagems to make it 'safe'. With the decriminalization of homosexuality, a special confusion enters the cultural scene as the category 'sexual preference' enters the political language. Sex is further removed from any natural definition. And when all this is set in the context of the global problems of overpopulation, another simple confidence is lost. Sex is no longer nature's continuing self-renewal, but the source of apparently destructive excess.

Yet, complication grows. It is often suggested that the attitudes that have occasioned the ecological crisis of our day are linked, in one way or another, to the repression of the feminine in male-oriented culture and to the oppression of women in the patriarchal structures of our history.[2] Consequently, a good deal of ecological reflection calls for a recovery of a sense of the nurturing, mothering, feminine reality of nature, as in the *Gaia* symbol. But this reconsideration of the feminine is occurring precisely at the time when the generative capacities of nature, and, more concretely, the maternal function of the actual women of our time are looked on with considerable misgiving.

The resultant ambiguity reaches its ultimate extreme in the complex issue of abortion. The incidence of legally terminated pregnancies ranges from one in five, to one in three, depending on the country considered. Not only is abortion regarded as a right—to be so freed from the constraints of nature—it has become, in the West at least, the emblematic issue of the feminist liberation. Both the character of such a movement and the kind of liberation it offered would doubtless have been different if its emblematic issue had been, say, world peace, or ecological harmony, or the economic recognition of the family. But it was not so; and while the wider concerns of women's liberation are in no sense vitiated, a polarizing clash of symbols has entered into the sexual identity of our culture: the elemental symbolism of the feminine as the bearer and nurturer of life; and abortion as symbolic of feminist liberation from the oppression of nature and culture.

Given the prevalence of legalized abortion in most countries today, the present generation, and those of the foreseeable future, grow up, or will grow up, with a sense of being survivors. Inevitably, sooner or later, it will dawn on them that they were amongst the eighty-to-sixty per cent who were allowed to be born; the rest of their generation was terminated. I do not see how we can evade a piercing question: what sense of life is being bred into such a generation? Perhaps they feel a strange kinship with endangered species. More to the point, what kind of ecological commitment can the world hope for from them?

Ecological conversation, busy no doubt with the intractable matter of overpopulation, is usually reticent on the abortion issue, even though a consistent pro-biotic ethic would seem to demand more open and thorough discussion.[3] Perhaps ecologists feel that there is enough polarization involved in their stance without exposing that particular nerve. But can the issue be indefinitely deferred? It does not seem a good preparation for a more tender and interactive relationship to nature as a whole when life at its most vulnerable is so intimately exposed to legally sanctioned extinction.

Through all this, the troubling area is the inner ecology of our culture. If there is no healing there, one must doubt our beneficent capacities in regard to the ecological wholeness of life. The issue was brought home when the first media reports of Romania's revolutionary break with a brutal regime hailed the reintroduction of abortion as a significant move in catching up with the West.

I am easily persuaded that people generally do not positively desire abortions; and that they are forced down this path by a deep violence in our culture. I suspect, too, that the most eager promoters of abortion on demand are themselves becoming appalled at their success. But somehow, that's how we are, and that's the direction society has taken.

In the meantime, the abortion issue remains one of extreme polarization. However, if a wiser comprehension of the issue cannot emerge, ecological concern will drift in ambiguity. It can easily become a frenzied overcompensation, driven by the guilt we dare not face. As long as abortion remains an unexamined 'given', it may be as well not to love our ecological neighbor as we love ourselves.

To return to the broader issue: human sexual relationships model a larger ecological relationship with nature as a whole. It seems to me that we cannot face the broader issue with any integrity without being prepared to address the particular. The profound confusion our culture is experiencing regarding sex, the most intimate relationship between men and women and their real or potential offspring, mirrors a larger ecological unease.

Paradoxically, we cut ourselves off from the great *eros* of life and existence by settling for the merely erotic. The passionate, or compassionate, range and depth of our immersion in the whole is blocked by fixation on sex in the narrowest sense. The essential dimension of the problem is perceptively expressed in the following words:

> ...Our relationship to our bodies, to our land, to our sexuality is singular, cut from the same model. We cohabit either lovingly or carelessly. And our erotic impulses are fully satisfied only when we are within an environment in which we are continually stimulated to care and enjoy. Eros is fully engaged only when we make the cosmic connection. Sexual love is both most passionate and most ordered when it assumes its rightful position within a nexus of erotic relationships that make up the natural world. Earthly love begins when we acknowledge our participation in an ecological bonding that joins all the species of life in a single commonwealth. Thus it is only when we deal with the dis-eased character of modern sexuality and the ecological crisis as *a single problem that is rooted in erotic disorder* that we can begin to

discover ways to heal our alienation from our bodies and from nature.[4]

The cultural problem of erotic disorder looms as the great challenge for the generations to come. I fear that their judgments will be harsh on the 'permissiveness' that masked the erotic incoherence of recent decades. With all the more reason, then, I will try to sketch the features of a more hopeful and wholesome tradition, even as it struggles to find more telling expression in today's world.

My remarks will deal with two limits of sexual experience: its sacredness—relationship to the transcendent; and its earthiness—relationship to nature.

2. Sex and sacramentality

With regard to the sacred character of sexuality, there is an explicit and complex tradition characteristic of Christian faith.

The classical expression here is the affirmation of marriage as a 'sacrament': a 'visible sign of invisible grace'. The experience of the divine is mediated through the very earthy, very bodily, very biological relationship of sex. Thus, the sexual relationship of the human couple is considered to embody the meaning and presence of the God who is Love (1 Jn 4:8). How much John's frame of reference extends to Christian marriage when he writes, 'Beloved, let us love one another, because love is from God; everyone who loves is born of God and knows God' (1 Jn 4:7), might well be doubted. Still, it is out of such essential Christian convictions that a profound theology of marriage can eventually emerge.[5]

The being-in-love of man and woman, in its intimacy and generativity, participates in the all-creative Being-in-Love that is the Trinity itself. Far from being a somewhat banal sphere of 'human relations', marriage is thus set, as an elemental relationship, within a matrix of cosmic and even ultimate relationship—of the divine persons to one another within the Trinity; of the divine to the human in Christ; of the natural to personal in the genesis of the human world.[6]

Whatever the ambiguity of sexual relationships and social strains on marriage today, the sacramental vision of marriage, and of sexuality in general, continues to be offered by Christian tradition as a focus of more inclusive connections, relationships and responsibility.

The present situation is, of course, far more complex than the above doctrinal description. For example, Paul Ricoeur, as one among many, has pointed to the historical complexity of our religious views on sexuality and marriage. This primal natural relationship has been subjected to a long process of de-sacralization, or secularization, if you prefer. Paradoxically, religious faith was a main cause of this: the One God of Israel, the transcendent God of call and promise, triumphed over the 'nature religions' of the *Baalim*, with their rituals of sacred sex. Faith in God left sex just sex, just part of God's good creation, not a sacred way of entry into union with the divine.[7] Similarly, the classical achievement of Greek philosophy, in its passage from *mythos* to *Logos*, was another influence in this secularizing process. The many gods of nature were banished in favor of the One Absolute Good.

Through such instances of religious and philosophical detachment from nature, the door was opened to a distorting orientation to creep into Western culture. Spirit, as opposed to nature and body, was the locus of union with the divine. Material nature ceased to be the dwelling place of God. For the higher *eros* of the spirit was distracted from its transcendent goal by physical passion.[8] Marriage, the just ordering of sexual instinct, had its place in the ethics and needs of temporal existence; but it belonged to the passing nature of this world; matter had nothing to say about what really mattered.

To leave such a diagnosis without qualification would mean settling for a caricature, valuable in dramatizing the distortion, but hardly presenting the complete picture. For religious faith, despite its sense of the transcendent character of God, even because of it, was preparing the ground for a more gracious sense of the immanence of divine within a holy creation. In the concrete reality of its history, two sets of attitudes were in conflict. On the one hand, there was the dark side of the sexual—its capacity to turn every man into a rapist, every woman into a harlot, and God into a demonic force of nature. Then, on the other, there was a contrasting sense of wholeness and even holiness in regard to the sexual. Biblical resources for a gracious vision of sexuality can be quickly mentioned.

First, the original vision of Genesis: God, in making human beings in the divine likeness, made them sexual: 'male and female he created them' (Gn 1:27). Secondly, the exuberant sexual imagery of the Song of Songs was classically evocative of God's relationship to his people: 'Set me as a seal upon your heart..., for love is as strong as death... many waters cannot quench love, neither can floods drown it...' (Cant 8:6f). A celebratory experience of sexual desire and ecstasy was revelatory of God's presence. Thirdly, the New Testament not only ratifies the integrity of marriage in the light of Genesis (Mt 19:4-7)—'what God has joined together, let no one separate'—but extends the sense of the radical goodness of sexuality as revelatory of God's love in its description of Christ's spousal relationship to the Church: 'Husbands, love your wives, just as Christ loved the Church and gave himself up for her...' (Ep 5:25–33): a 'great mystery' indeed (v.32).

Out of such resources develops the intense, compact affirmation of marriage as a sacrament in the way described above. Yet, there is every reason to feel that we are only at the beginning of an engaging theology of marriage. The challenge that emerges is how to combine a deeper sense of participating in the mystery of creative Love with an enjoyment of interpersonal intimacy, even if it means to be embodied in the 'labor pains' of the whole of creation (Rm 8:22).

A renewed sacralization of sexuality has to struggle against an inadequate symbolization of God's transcendence, when this is interpreted as an indefinite removal from his sexual creation. Christian tradition knows many a caricature of God as the transcendent 'Father', blessing the human sexual union merely as part of his plan—presumably, to get more souls for heaven!—and in the process providing some decently regulated consolation for his human children, as a reward for their dedication to the generative task. But in this version of things, a kind of unbiblical spiritualism seeps into our understanding of sexuality. Sexual relationality is left awkwardly extrinsic to that trinitarian relationality which is the matrix of all interpersonal and natural relationships.

In short, sexuality drifts to the periphery of the field of the Spirit's action throughout creation. It is the odd dimension in the 'Christogenesis' of the cosmos. What is called for, in an enlarged Christian vision, is a new appreciation of how the sacramentality of marriage (and, proportionately, of all sexual relationships characterized by mutual fidelity and generativity) might be more properly located in the couple's experience of the Holy Spirit.

The holy and whole-making Spirit is the field of self-transcending love pervading the whole becoming of the cosmos. Marriage is born out of the call to join, to participate in, what is coming to be, to share in the engendering of the great communion of life in Christ. Marriage is not so much made 'in heaven'; but made in response to the universal ecstasy of life toward its fullness and wholeness of communion. It is a couple's entry into the 'unity of the Holy Spirit'. In the Spirit, they become generative members of the Body of Christ. In Christ and his Spirit, they celebrate a certain homecoming of creation in the mystery that unites them.[9]

Christian marriage, then, is not best seen as a structure imposed on the couple from above, 'made in heaven'. More earthily and experientially, it is a union inspired from within—the Spirit acting from the within the earth, from within all its generations; from within its search for communion and its desire for fulfillment. Here the couple says 'we' in a way that is founded in the larger 'we' of creation; and which finds its deepest promise in the way the trinitarian community says 'we', in the unity of the Holy Spirit. It is a participation in the 'we' of the Holy Wholeness of all things in Christ.

3. The desire of creation

Admittedly, unless a deeper appreciation of the revelatory character of sexual experience is registered, all that sounds like so much more abstract theology. Still, I think there is evidence of a growing wisdom.[10] The current problem of sexual relationship is to understand it in a universe of deeper and broader connections: to link it both into the deeper eros of creation generally, and also into that fundamental 'desire' for ultimate belonging which is distinctively human. The crucial point is this: it is not a matter of having less desire in our lives, nor of reducing such desire to the narrowly sexual. It is rather a question of earthing the sexual in a larger life of desire and relationship. Here the individual ego must give itself into, die into, a deeper, broader self-realization in the universe.

If everything is reduced to sex, the totality of desire has nowhere to go. Yet, if sex is left out of desire, then desire is disembodied. Consciousness oscillates between an apathy in regard to its transcendent connection, and an obsession with regard to the genital. Hence, the challenge: to liberate the deep *eros* of our existence into its true proportions. On this point, the following words of Simone Weil are arresting:

> If people were told: what makes carnal desire imperious in you is not its pure carnal element. It is the fact that you put into it the essential part of yourself—the need for unity, the need for God—they wouldn't believe it. To them it seems obvious that this quality of imperious need belongs to the carnal desire as such. In the same way it seems obvious to the miser that the quality of the desirability of the gold belongs to the gold as such, not to its exchange value.[11]

Sexual desire is, then, often made to carry the whole passion, the total *eros* of being. In the process, both are diminished. For Christian faith, the ultimate symbol of self-realization in the universe is not the orgasm—sexual desire, fulfilled and spent—but the cross and resurrection of Jesus, the death to the ego-self for the sake of a life of full relationality in the Spirit. His death is born out of a desire impatient of anything less than 'life to the full'. Admittedly, sex and the cross of Christ are strange bedfellows, but that may be exactly the consideration we most need. This is to suggest that the mystics and prophets of the world are, in reality, the most *eros*-filled of human beings—men and women of untrammeled desire.

The deepest *eros* of our lives, what we really want and have always wanted, remains unsatisfied with anything less than complete self-realization. The dynamics of consciousness reach beyond the successive ego-structures of provisional existence, to the mystery which is the home of our being. The following quotation from Sebastian Moore underscores this point, and leads to the central issue:

> Desire is love trying to happen. It is the love that permeates all the universe, trying to happen in me. It draws into its fulfilling meanings all the appetites of our physical being. It turns the need for shelter into the sacrament which is a house. It turns the need for food and drink into—well, *Babette's Feast*! And it turns sexual passion into—ah, there we have a problem... The sentence ends with 'marriage', and this is true. But the biggest ethical gap in the Christian tradition is the failure to say anything much as to *how* that taking up of sexual passion into authentic desire, of which marriage is the institutionalizing, happens.[12]

The communion wafer is hardly recognizable as appetizing bread. Sickly sweet sacristy wine falls far short of even the 'rough red' of our ordinary consumption—to say nothing of the delight of a more festive vintage draft. If the sacramentality of the eucharist suffers from such artificiality, so, too, does the sacramentality of marriage artificially constructed on a kind of fleshless notion of sex. On the one hand, the shadow language of our usual colloquialism is too improper for public discourse. On the other, such language as we have is either so personal or so private or so sacred or so clinical. The confused focus of our sexual expression is a symptom of a deeper oddness: unease with the place of our sexual existence in the integrity of creation.

No doubt, any reclamation of the sexual self often masks the subtle play of egoism. On the cultural level, it can play into the hands of an imperious, erotic incoherence. And, of course, the Church's moralizing often mirrors such confusion. By offering guidance on sexual morality while leaving the fundamental issues unexamined, it acts more like a chaplain to the incurable than as the bearer of life's greatest invitation.

The manner in which Christian moral doctrine has both affected, and been affected by, the bias of Western culture is a long and complicated story, some elements of which I have touched on above.[13] Embedded in every context of discussion is the position of Stoic philosophy on pleasure. Pleasure was not lawful in itself, but only as an adjunct to the worthy end, in this case, of procreation. The conclusion drawn, especially when you allow for the different forms of distrust of the body as they have appeared in Manichaeism and Neo-Platonic thought, was that sexual enjoyment had no intrinsic value; it had to be legitimated by its biological end. Hence the logical impossibility of allowing for marital intimacy as an end in itself. The activity of sex, the intimacy of sexual relationship, was trapped as it were in a single justification: the value of procreation.[14]

Between understanding sex merely as recreation or only as a means of procreation lies the great and perplexing gulf that we have been exploring. A significant marker in the development of Christian thought (and feeling) is Vatican II's treatment of married love.[15] Authors now speak of a paradigm shift from a primarily biological and juridical notion to one that is more interpersonal, spiritual and existential—a new hierarchy of values. I suspect that an infusion of ecological and cosmic values will need to be the next step if such a paradigm is to be productive. Such development as has occurred could result in a further uprootedness of marriage and sexuality from its ecological and cosmic connection. As the undifferentiated 'biologism' of earlier accounts is rejected, 'nature' is left as an increasingly obscure underpinning, transcended by the interpersonal.[16] It would be tragic if a new theology of marriage and sexuality was really only apprenticing itself to a de-natured culture. A bleak prospect but not a necessary one.

Interestingly, John Paul II devoted his weekly audiences from November 1979 to January 1980 on 'the nuptial meaning of the body' disclosed in the second and third chapters of Genesis.[17] Though he spoke as no other pope had ever done on the positive meaning of sexual existence, the stark fact of 'The Fall' of our first parents had eventually to be faced: cosmic disharmony, occasioned by the human rejection of creator, enters creation; and the psychosomatic relatedness of our being in the presence of God is skewed by the presence of the demon lust—sex uprooted from its original place in God's universe.

Despite his appreciation of the pope's profound reflections, Sebastian Moore has insisted on taking them further.[18] Whilst we cannot expect the pope to be an ecclesiastical D. H. Lawrence, we are uncomfortably aware that even the best doctrinal treatment of sex is hampered by considerable abstractness. For even despite the Fall, sexual relationships continue to be somehow sacramental and revelatory. Granting the disharmony that pervades our experience, is it entirely fair to interpret such tensions primarily as the failure of the lower instincts to obey the higher, the spiritual, the soul-values of our existence? It could be that the incoherence of our relational existence has a deeper cause: the failure of these 'higher' levels to befriend and learn from the lower.

After all, in the presence of God, our first parents, the 'would-be gods' of the Genesis story, come to feel shame at their nakedness, their animality, the sexuality inscribed into their nature, and into nature as a whole. They feel the consequences of trying to escape out of their sexual creatureliness into a 'spiritual', godlike sphere. The result is that what they have disowned now reasserts itself as 'lust'—sexual desire with nowhere to go; sexual desire uprooted from the whole mystery of life; sexual desire violently de-natured, out of harmony with the natural world of God's creation.

Thus, 'original sin' is a tragic split in the original mystery of the relational selves we were meant to be. Grace comes as a healing reconnection of the spiritual with the sexual in our consciousness. It is the gift that restores nature by making us participants in a larger cosmic whole, by putting our souls back into our bodies, and putting our bodies back into the great process of nature itself; finally, by setting nature as whole, in a universe of divine creation and connection.

Such glimmerings, however tentative in formulation, do lead to real questions. Whatever answers may eventuate, discussion has to be enriched by asking: not what does the God who 'is love' reveal about sex? but what does good sex tell us about the love of God?; not how can the sexual union of man and woman be said to be one of the sacraments? but what does wholesome sexuality tell us about what is sacramental?; not how does the couple transcend nature in their 'interpersonal relationship', 'spirituality', and existential experience? but how does the universe, as a field of connections and relationships, draw such a couple to be embodied in its emerging wholeness?; not how can unbridled passion be controlled?, but how can jaded, fixated passion be released into the dimensions of true desire?; finally, not what does natural law teach us about the regulation of sex, but how does participation in a sexual universe teach us concretely what that law should be?

Either way, it hurts: to accept a world of unredeemed sexuality and to make moralizing gestures toward it; or to accept the experience of good sexuality, as it occurs, and go with what it reveals—and demands. Death to the ego in the presence of the other—of spouse, child, the generations, the ecology of the biosphere—and in final surrender to the mystery intimated in it all, is what we all rightly perceive as disproportionately difficult. 'Nature' should be a lot more 'natural', we feel. At times, it is: the joy and ecstasy of sexual love... But falling in love eventually demands that we stand in it, and follow it through, to whatever end is revealed. In such a context, everything theology might say about the sacramentality and the spirituality of marriage can be heard with new perception.

There are many particular issues that are beyond both my competence and the limited scope of this brief reflection. Nonetheless, I would like to offer short remarks on two further points.

For instance, at one extreme, I have not mentioned chastity, let alone celibacy, or virginity. With reason. Such words lead quickly into embarrassed silences. The public supposition is that the sexual instinct is ungovernable.[19] The best that can be offered is damage-control—'safe sex' and so forth. We have to suspect a disconcerting absence of values at this point, perhaps even the occurrence of some kind of deep mutilation in the moral tradition of humanity. Still, other forms of vigorous and even terrifying renunciation are in fact being demanded (and even imposed) in the name of higher social values of one kind or another. Note the variety and intensity of campaigns against nicotine, alcohol, drugs, overeating. The asceticism of 'fitness' can be so extreme as to cause concern on the part of medical authorities. But most of all, the wholesale change of lifestyle foreshadowed by ecological concerns presumes the desirability of some enormous redefinition of what was previously non-negotiable in regard to the simple (or artificial) pleasures of life!

What, then, of the values of chastity (the moderation of sexual activity), of celibacy (a chosen form of abstention from sexual activity), and virginity? The derisory manner in which such longstanding and widespread values[20] are regarded looks like a classic case of *ressentiment*, as Scheler and Lonergan employ the term. Such a form of disaffection amounts to a continual feeling of hostile reaction to a scale of values which one desperately needs, but feels unable to implement. The value in question is continuously belittled; it is turned into a non-value, something inhuman, for it represents a threatening challenge. Note the way the ecologically committed are often dismissed as mere 'Greenies'; the way activists for human rights are felt to be interfering busybodies; the way a critical feminism can be regarded as strident nonsense; the way Aboriginal sacred sites are located in the realm of the absurd.

The counter-cultural, the effective questioning of the destructiveness of a culture, is deeply troubling. We all feel defensive in the face of the wholesomely subversive. What a society most needs, hurts, unsettles, and makes 'impossible' demands. In regard to *ressentiment* in general, Lonergan writes:

> perhaps its worst feature is that its rejection of one value involves a distortion of a whole scale of values and that this distortion can spread through a whole social class, a whole people, a whole epoch. So the analysis of *ressentiment* can turn out to be a tool of ethical, social and historical criticism.[21]

It seems to me that this is the case in regard to chastity and its associated values and lifestyles. Given the sexual misery of our age, chastity is the value we most need, but feel most incapable of realizing or expressing. How much the demeaning of such a value has resulted in a distortion of the whole ecology of human values, how much that distortion has spread through whole social classes, or peoples or the epoch itself, would be a matter of some debate. Still, I doubt that many would disagree that there is a problem of considerable magnitude; and that the situation of sexual incoherence I briefly described in the first section of this chapter is real enough.

On the other hand, by raising the question of sexuality within an ecological and cosmic context, we are confronted with the intimate challenge of 'ordering' with fresh imagination, passion, and, indeed, discipline the 'disordered' sexual energies of our culture. If the erotic fixation of our culture presents us merely with a world of sex-objects, a fresh sexual imagining of our world would unfold in a universe enticing us out of ourselves into communion and self-surrender. In each particular relationship of sexual love there is a sacramental implication of a larger mystery of universal love, inviting us to join.

Likewise, we are being challenged to a new degree of sexual passion, something that consumer-sex so little tolerates with its range of disposable sex-objects, its fixation on performance and its cult of the orgasm. Sexual passion, in this broader perspective, is the sense of being embodied in the passion of all being for life and wholeness, a participation in the ultimate joy-in-being out of which the universe arises.

The requirement of discipline is, I suppose, the most obvious, to integrate the sexual into self-transcendence, as other-regarding, life-promoting love. Far from being repressive, it is a matter of welcoming and befriending sexual energies as sacramental, holy—healing and whole-making—creation's insistent summons into a larger body of connections, into a universe of relationship and communion.

Still, some kind of new asceticism is a necessity. A paradoxical witness in this connection is the variety of celibate lifestyles as they have appeared in Eastern and Western forms.[22] As an element in the ecology of our sexual existence, it is important to understand religious or ascetic celibacy, not as a rejection of sexuality, nor as demeaning marriage, but as dramatizing a transcendent dimension latent in all sexual relationships.[23]

In its free renunciation of exclusive, genital relationship to the other, such celibacy still participates in the desire for ultimate union that flares through the universe. As it concentrates on the transcendent meaning of human existence, it anticipates the final intimacy that each one must face—in death. As such, celibate witness is exploring something on behalf of all. Within the ecology of our sexual nature, the necessarily rare celibate vocation intends and, in a sense, secures ranges of value that pervade all relationship: self-transcendence, generosity, a sense of the mystery of the other beyond one's own needs, the discipline inherent in all mature relationships, self-sacrifice for a more inclusive community.[24]

Though I have not intended to give a disquisition on religious life, the value of celibate witness should, I believe, be noted. By contesting the eroticism of our culture, it invites us all to a deeper appropriation of sexuality as a mystery of relationship, and as a relationship to the universal mystery itself.[25]

4. A developing tradition

I have not directly faced the contraception issue, because I do not think it is best directly faced. It is about something else: how are actual human beings related to the actual, concrete good of 'nature' or the biosphere generally?

The issues are of extreme complexity and the literature inexhaustible. Hence, I will limit myself to a brief comment.

The theology of marriage can hardly be separated from the Catholic moral tradition associated with it. In such a theological tradition, a perception of the 'natural' structure of sexuality was fundamental. It was felt that the meaning of marriage is obfuscated if the generativity of sexual relationship is unnaturally blocked. Human sexuality has to respect the rhythms of life and not be distorted by any artificial manipulation. Now such a position seems like an impractical dream. In the face of overpopulation problems, a more personalist understanding of marriage, and the economic and psychological pressures of modern life, the cost of maintaining such a tradition has been extreme. The debate ensuing on the publication of *Humanae Vitae* in 1968 continues.[26]

One thing is worth noticing: the similarity between arguments against this official Catholic position and those used against environmentalists struggling to maintain the ecosystems of a rain forest or a reef. In both cases, the antagonists are quick to subordinate 'nature' somewhat hurriedly to other values, and to ostensibly more realistic concerns.

Admittedly, there are difficulties with an idealization of 'nature'. The basic problem, it seems to me, is that Catholic thought, as well as common forms of ecological doctrine, have both tended to relate the human 'organically' rather than transformatively to 'nature'.[27] How nature might be subsumed into a higher integrity through human action tends to be ignored. But before saying more about that, we note the profound and challenging issue struggling for expression in the hard teaching to which we have referred. In trying to express how human love should be embodied, Catholic tradition was implying something terribly important about our relationship to nature and its processes. Even while accenting the interpersonal dimension of marriage, even as it allowed for 'natural family planning', Catholic theology has usually insisted on placing sexual activity in the context of 'natural law'. Conscience was more than a private spiritual matter;

it had to include a consciousness of biological integrity which, on its own level, was to be inviolable. Human sexual intimacy had to embody a larger intimacy with the given biological processes of the natural world. What contemporary culture sees as most private and personal, the Catholic tradition places in the dynamic of cosmic participation in which the rhythms of generativity and the organic structure of our embodiment must be respected.

The cost of implementing such a vision has appeared disproportionately high: the demographic explosion, the economic pressures, the sexual confusion of modern culture hardly permit a calm reception of such a doctrine, let alone a ready implementation of it. Still, it seems to me, there is something struggling for expression here of deep ecological importance. It is this: in the measure we can slowly learn to relate to one another in human intimacy in a way that respects the biological integrity of the other, we begin to express a model for understanding our relationship to nature generally. Our sexual relationships are an expression of the kind of ecological commitment we are in fact choosing to live. The words of Wendell Berry, a great ecologist who owns no special responsibility to the Catholic position, express this point with vigor:

> There is an uncanny *resemblance* between our behavior toward each other and our behavior toward the earth. Between our relation to our own sexuality and our relation to the reproductivity of the earth, the resemblance is plain and strong and apparently inescapable. By some connection that we do not recognize, the willingness to exploit one becomes the willingness to exploit the other. The conditions and means of exploitation are similar. The modern failure of marriage that has so estranged the sexes from each other seems analogous to the 'social mobility' that has estranged us from our land, and the two are historically parallel. It may even be argued that these two estrangements are very close to being one, both of them having been caused by the disintegration of the household, which was the formal bond between marriage and earth, between human sexuality and its sources in the sexuality of creation.[28]

The emerging challenge has to do with respecting sexual integrity as an indicator of an inestimably greater challenge, that of respecting the integrity of nature as a whole, and of the human place within it.

Any attempt to be more specific soon reveals a hive of complexities. I am not sure the present limited reflection needs to stir up an angry swarm. Nonetheless, I make the point: the Catholic tradition of a sacramental and ethical understanding of sexuality is an improbable source of ecological responsibility as a radical demand. Sex and ecology are seldom brought together in any discussion, possibly because we dread where it would lead.

At first glance, at least, we veer close to a contradiction if we think we can be thorough-going ecologists while manipulating the human organism and human biology in any way we wish. Such biological interference may unmask a strange exclusion of human sexuality from the realm of the natural, perhaps because one may feel that human population is the main danger to ecological well-being. On the other hand, we could be all for the integrity of human sexuality, yet ignore the biosphere sustaining human and other life; and so refuse the specific role of human creativity within the realm of nature. The more philosophically inclined will point out that both concerns can only be resolved in a much deeper and far more embracing notion of how nature and the human are related—perhaps more in terms of the artist working within the limits and possibilities of a given medium.

What happened in such a formulation of the Catholic doctrine on the integrity of sexual nature was that the neuralgic point of something far more vast was being touched: how are we to relate to the biosphere generally? In defying the canons of a contraceptive culture, Catholic teaching in this area has sounded like a voice from the past. The unsettling possibility lies in it being a message from the future, when the sexual, the sacramental and the natural come together in new integrity. Not to mention that possibility is, in my view, to foreclose on the possibilities of imagining the world otherwise. To speak of 'natural' family planning, and the consequent ban on contraception, makes the powerful point that I have outlined above. Sexual relationship models our relationship to nature generally. It is governed by 'natural law'. It demands the asceticism of wholeness and integrity; a disciplined affirmation of the value of life and its processes. This is a precious insight.

What remains a problem, as I mentioned above, is the philosophical and physicalist notion of nature embedded in current ecclesiastical formulations. In the tradition stemming from Aquinas, Natural Law is not something imposed from the outside, as it were, but enters moral consciousness through 'right reason', i.e., in the use of the best resources of human intelligence; and more deeply and tacitly, through 'connaturality', i.e., an affective inclination toward the whole, concrete human good. To that degree, Natural Law occurs within the reflections of the human mind and in the aspirations of the human heart. It is not something simply objectively inscribed in the physical or biological world, even though the structures and processes of that world are a value to be respected.

Now, it is exactly here that new considerations have emerged. As a more personalist approach to the sexuality emphasizes the unitive aspect of sexual relationship, it points to the natural foundation of this aspect in an empirical notion of nature. Where before 'nature' was a philosophical notion revealed in the essentially procreative structure of sexual organs, biological science reveals the nature of human intercourse as one characterized by randomness in its generative outcomes: most such acts are infertile. There is no metaphysical Natural Law linking intercourse to conception, as Natural Family Planning knows.

When a metaphysically constructed Natural Law yields to one derived from 'right reason', and when such 'right reason' is further illumined by data from the biological sciences, a new context emerges. And the discussion goes on, not a little obscured by 'natural law' being used in at least three different senses! How then does the human couple live out its connection to, and within, the biological process? How does such a couple subsume into its relationship the generative potential of intercourse? Natural Family Planning represents one sophisticated human intervention. But does that represent the full range of the transformative relationship with nature—nature as it comes to be in human exploration—and subsumed into the service of the actual human good? What does 'artificial' contraception mean? What is truly 'natural', especially in light of current economic and demographic pressures?

Once the notion of 'nature' begins to operate outside an undifferentiated religious or philosophical context, what was accepted as natural to the 'right reason' of one era appears impossibly unnatural to the 'right reason' of another. A new discernment of the will of God acting in all creation, and of a divine wisdom calling human history into express collaboration with the Lord and Giver of life, becomes necessary.

These general remarks, deferring as they must to the more thorough analyses of Ethics and Moral Theology, must include one further consideration. Again it is a matter of utmost concreteness—and urgency. We are being invited, I feel, to consider human fertility as set within an actual context: that of the limits and needs of an empirically understood natural world.[29] Human generativity must serve life as a global value. It has an ecological responsibility. Respect for the biological process as it has been instanced in the Catholic tradition of sexual morality must now make a quantum leap into a more comprehensive commitment to the global biological process, into the realm of nature under severe ecological stress.

A 'pro-life' commitment needs now to unfold into concern for the ecological well-being of the planet, as well as for the future well-being of the generations to come. Here, above all, Natural Law is being revealed in new responsibilities and as suggesting new limits. World population is predicted to double in the next fifty years. Naturally—I use the word advisedly—the human race has to explore other dimensions of generativity: there are more ways of inhabiting the planet than by populating it to death. We are in a new historical situation, and one would be singularly lacking in wisdom to offer simple solutions. The necessary change in lifestyles and scales of values might well be immense. The only thing that is clear is that human history is being called to a momentous new responsibility.

Notwithstanding the limits and opportunities that the law of our actual historical nature might suggest, the human sexual relationship remains as a participation in the whole mystery of life. This is, as I see it, the enduring value of the Christian tradition to which I have referred.

In the complex history of its contestation of all views which are either contrary to life, or reductive of sexuality to mere gratification, such a tradition has consistently opted for the essentially self-transcending dynamism of sexual existence. *Agape* meets *Eros*, not to belittle or contradict it, but to make of our sexuality a 'life principle', generating and nourishing life, in all its connections. Radically, through their sexual coexistence, men and women participate in the energy of the Spirit as the life-giving and all-connecting mystery working through all creation. Such a sacramental vision of sexuality is in sharp contrast with *Thanatos*, 'the death principle' of dissolution and self-absorption which so deeply marks the erotic vertigo of our culture.[30]

In its unitive and generative aspects, sex, in a developing Christian tradition, continues to be affirmed as sacred, a sacrament of God's healing and renewing presence. By reclaiming our sexual nature in a more wholesome way, the future can be graciously embodied in our generation—not just physically, but in all the resources of intelligence, love and care which we can bring to bear.

1. From the frontispiece of Sam Keen, *The Passionate Life: Stage of Loving*, Harper and Row, New York, 1983.
2. Anne Primavesi, *From Apocalypse to Genesis: Ecology, Feminism and Christianity*, pp. 24–61 makes a good case, without, however, addressing the emblematic importance of the abortion issue.
3. A significant exception is Albert LaChance, *Greenspirit. Twelve Steps in Ecological Spirituality*, pp. 95, 106, 109, 120f. This excellent practical book, addressed to all religious traditions, faces fundamental issues with great courage and insight.
4. Sam Keen, *The Passionate Life: Stages of Loving*, pp. 231f.
5. For the fundamental context, see my *Trinity of Love*, pp. 142–64.
6. For a general remark on this point, see Catherine La Cugna, *God For Us: The Trinity and Christian Life*, Harper, San Francisco, 1991, pp. 406–08.
7. Paul Ricoeur, in H. Ruitenbeeck (ed.), *Sexual Identity*, Dell, New York, 1970, pp. 13–24.
8. For an abundance of interesting material, see Peter Brown, *The Body and Soci-*

ety: Men, Women and Sexual Renunciation in Early Christianity, Columbia University Press, New York, 1988.

9. For a more 'Spirit-ual' understanding of marriage, see Heribert Mühlen, *Entsacralisierung*, Shoeningh, Paderborn, 1970, pp. 473–505.
10. As a good example, see the provocative and insightful remarks of Sebastian Moore, *Jesus the Liberator of Desire*, Crossroad, New York, 1989, pp. 80–107. In what follows, my indebtedness to this work is as obvious as it is grateful.
11. From Simone Weil, *First and Last Notebooks*, as also quoted on the frontispiece of Sam Keen, *The Passionate Life*.
12. Sebastian Moore, *Jesus The Liberator of Desire*, p. 93.
13. Ibid., pp. 94–100.
14. For references and further details here see the very comprehensive Theodore Mackin, *The Marital Sacrament: Marriage in the Catholic Church*, Paulist Press, Mahwah, New Jersey, 1989, pp. 95–7.
15. See David Thomas, 'Marriage', in *The New Dictionary of Theology*, Joseph Komonchak et al (eds.), Michael Glazier, Wilmington, Delaware, 1987, pp. 624–8, and Vincent Genovesi, 'Sexuality', pp. 947–54.
16. Note how the theology of marriage tends to develop away from its natural location even in the document that most wishes to respect such integrity: '...marriage is not then the effect of chance or the product of evolution or of unconscious forces; it is the wise institution of the Creator to realize in humankind his design of love' (*Humanae Vitae*, par. 8.). The concern of these reflections is to explore how the 'design of love' is indeed immanent in the Spirit-guided evolutionary dynamics of nature. For a general comment, B.V. Johnstone, 'From Physicalism to Personalism', in *Studia Moralia* XXX/1, 1992, pp. 71–96.
17. See Mary Durkin (ed.), *Feast of Love: Pope John Paul II on Human Intimacy*, Loyola University Press, Chicago, 1986. See also T. Mackin, *The Marital Sacrament*, pp. 554–5.
18. Sebastian Moore, *Jesus The Liberator of Desire*, pp. 95–107.
19. Illuminating information on the ideals and practices of another and a far distant culture is found in Peter Brown. *The Body and Society: Men, Women and Sexual Renunciation in Early Christianity*.
20. For a larger context of discussion, not only in Christianity, but also including the experience of Hinduism, Buddhism, Judaism and Islam, see Diarmid O'Murchu, *Religious Life: A Prophetic Vision*, Ave Maria Press, Notre Dame, Indiana, 1991, pp. 118–41.
21. Bernard Lonergan, *Method in Theology*, Darton, Longman and Todd, London, 1971, pp. 33; 273.
22. For a most perceptive treatment of this issue, see Diarmid O'Murchu, *Religious Life: A Prophetic Vision*, pp. 14–59; 118–41.
23. Ibid., pp. 193–211.
24. On this subject, see the section 'Asceticism and Chastity Required by Life' in Thomas King, S.J., *Teilhard de Chardin*, Michael Glazier, Wilmington, Delaware, 1988, pp. 158–74.
25. D. O'Murchu, *Religious Life: A Prophetic Vision*, pp. 227–41.
26. See Janet E. Smith, *Humanae Vitae: A Generation Later*, Catholic University of America Press, Washington, 1991.
27. Here I refer the reader back to 'A Second Circle of Connections', Chapter 6, 'Models of ecological action'.
28. Wendell Berry, in 'The Body and the Earth', in *Recollected Essays, 1965–1980*, North Point Press, San Francisco, 1981, pp. 304. For this quote and further remarks see Murphy, *At Home on Earth: Foundations for a Catholic Ethic of the Environment*, pp. 12–15.
29. See Sebastian Moore, 'Ratzinger's Nature isn't Natural. Aquinas, Contraception and Statistics', in *Commonweal* 26 January 1990, pp. 49–52.
30. I know the use of these terms differs from Freud's original employment of them in *Beyond the Pleasure Principle* in 1920, in which the sexual instinct was, of course, located on the side of *eros*. But what would such an uncompromising realist have made of the erotomania of our day when sexual liberation, without any correlative spiritual liberation, has become mired in a sexual licentiousness?

A Concluding Remark

The theological enterprise of a faith seeking understanding has been presented as faith making new connections. No doubt the variety and depth of such connections can be indefinitely extended in a new collaboration of faith and science and art and all forms of human learning. For instance, apart from its fragmentary familiarity with burgeoning areas of modern science, one obvious limitation of this present treatment is its scant reference to the interfaith context of today's global search for meaning. To that degree, I leave you with something of a prelude. But at least my effort here is a point of departure, of further learning and search; and splendid resources abound.[1] On the other hand, because it suggests the planetary and cosmic dimensions of the focal mysteries of Christian faith, because it evokes the outreach of consciousness illumined by such a sense of reality, it is a good point of departure, a wholesome beginning: there are no essential limits to the faith, hope and love that, in the foregoing pages, we have attempted to express.

Our spiral of connections took us first of all to consider theology as the expression of a connective faith. In the process, we noted both the connective mission of the Church, and the implicit inclusiveness of Christian experience. There were great basic simplicities from which to begin, even if the complexities were daunting.

Secondly, we took in a variety of contexts in which these implications were being invited to become explicit. The 'whole story', the search for a new wisdom through a new paradigm, in a 'new age', the different models of ecological interaction, provoked a re-examination of the Christian experience of God's love, as it kept on being love in the context of connections.

Thirdly, the spiral of our reflections connected within the universal meaning of the Word Incarnate. Here the question emerged as how to 'word' or express the mystery of the Incarnation as the great poem of universal transformation?

Then followed, as a fourth circle, the connection of all existence within the mystery of creation, freshly experienced in the wonders that science discloses.

Intimately related to the mystery of creation was the focus of a fifth arc in the spiral of connections, creation conscious of itself in human existence. There is an outreach and relationality in human consciousness. The dynamic integrity of the human occurs within the universal process, not outside it. Our existence is earthed in the well-being of the planet.

The trinitarian mystery suggested a range of ultimate connections. The universal process is grounded in the processive vitality of the divine. Created reality is relational as an emerging image of the originative reality of trinitarian love. Thus, the sixth circle of connections.

The seventh earths the experience of faith in its most intense sacramental expression, as an intimation of a eucharistic universe. Against such a background, an ecological commitment is a moral necessity.

A concluding section deals with the two dimensions of death and love. The spiral eddies, first of all, in the question of death. Is our whole emerging existence bent on self-extinction, or a movement into further transformation? We conclude the grain of wheat does fall into fertile ground.

With sexual love, old mysteries and new problems are present. Here, too, a depth of the sacramental is found, and a connection with great unitive *eros* of creation. Yet, in the strained ecology of our planetary existence, we are called to a fresh moral reappraisal of the 'natural' meaning of the sexual connection.

In a very real way, Teilhard de Chardin provides the conclusion to which all these reflections were headed. In his famous meditation in *The Divine Milieu*,[2] he communicates a mood familiar to most of us today as Christian faith tries to get new bearings for the long way ahead. As he descends from 'the zone of everyday occupations and relationships where everything seems clear', into that 'inmost self', to 'that deep abyss whence I feel my power of action emanates', he reports a profound vertigo. He feels he is losing contact with the self of routine relationships, to the point that:

> At each step of the descent, a new person was disclosed within me of whose name I was no longer sure, and who no longer obeyed me. And when I had to stop my exploration because the path faded from beneath my steps, I found a bottomless abyss at my feet, and out of it came, arising from I know not where, the current I dare call *my* life.[3]

We have been reflecting on how the uncanny occurrence of life and existence is coming home to human consciousness in new and often overwhelming ways. The settled identities structured for us in the history and geography, the culture and the philosophy, the religion and science of former days are being dismantled. We are becoming vulnerable to the mystery of it all again, and awakening to the necessity of finding a new place within what both transcends and enfolds us. It is an unsettling experience to find oneself on the brink of the unfathomable. So much so that Teilhard confesses:

> I then wanted to return to the light of day and to forget the disturbing enigma in the comforting surroundings of familiar things—to begin living again at the surface without imprudently plumbing the depths of the abyss.[4]

But neither looking for solace in the old connections, nor throwing oneself with greater energy into daily work, nor clinging simply to the old certitudes protects us for long. We cannot hide:

> But then beneath this very spectacle of the turmoil of life, there reappeared before my new-opened eyes, the unknown that I wanted to escape. This time it was not hiding in the bottom of the abyss; it disguised its presence in the innumerable strands which form the web of chance, the very stuff of which the universe and my own small individuality are woven. Yet it was the same mystery without a doubt: I recognized it.[5]

The mystery hidden in the depths of our experience emerges in the length and breadth of the history that has brought us forth. We live in a world of connections. Our existence is woven into the fabric of a vast, chancy coexistence. The sense of such intimate dependence on a totality, of being at the mercy of such an intricate network of improbabilities, brings its own dizziness:

> Our mind is disturbed when we try to plumb the depth of the world beneath us. But it reels still more when we try to number the favorable chances which must coincide at every moment if the least of living things is to survive and to succeed in its enterprises. After the consciousness of being something other and something greater than myself—a second thing made me dizzy: namely the supreme improbability, the tremendous unlikelihood of finding myself existing in the heart of the world which has survived and succeeded in being a world.[6]

The fifteen billion years of cosmic emergence, the miracle of life that has occurred on this tiny planet, have given each of us, as a supreme improbability, to ourselves; and to one another. Our existence becomes a calling, to relate to mystery which has given us into being, and to make connections of care with everyone and everything that is already part of our identity. There is a kind of distress evident in such belonging. Yet a bewildered consciousness can find a healing, and grow to hope, in the presence of the mystery which has given itself into the heart of the universe:

> At that moment, as anyone will find who cares to make this same interior experiment, I felt the distress characteristic of a particle adrift in the universe, the distress which makes human wills founder daily under the crushing number of living things and of stars. And if something saved me, it was hearing the voice of the Gospel, guaranteed by divine successes, speaking to me from the depths of the night, 'It is I. Be not afraid'.[7]

1. For example, Ninian Smart and Stephen Constantine, *Christian Systematic Theology in a World Context*, Marshall Pickering, London, 1991.
2. P. Teilhard de Chardin, *The Divine Milieu*, trans. unnamed, Harper Torch, New York, pp. 76–78.
3. Ibid., p. 77.
4. Ibid., p. 78.
5. Ibid., p. 77.
6. Ibid., p. 78.
7. Ibid., p. 78.

BIBLIOGRAPHY

Anderson, Bernard W. *Creation in the Old Testament.* SPCK, London, 1984.

Appleyard, Bryan. *Understanding the Present: Science and the Soul of Modern Man.* Picador, London, 1992.

Barbour, Ian G. *Religion in an Age of Science.* SCM Press, London, 1990.

Bednarowski, Mary Farrell. 'Literature of the New Age: A Review of Representative Sources'. *Religious Studies Review* 17/3, July 1991, pp. 209–12.

Berry, Thomas. *The Dream of the Earth.* Sierra Club Books, San Francisco, 1988.

Birch, Charles. *On Purpose.* NSW University Press, Sydney, 1990.

Bookchin, Murray. *Toward an Ecological Society.* Black Rose, Montreal, 1980.

Remaking Society. Black Rose, Montreal, 1989.

The Philosophy of Social Ecology. Black Rose, Montreal, 1990.

Bouyer, Louis. *Cosmos: The World as the Glory of God.* trans., Pierre de Fontnouvelle, St Bede's Publications, Petersham, Massachusetts, 1988.

Bradley, Ian. *God is Green. Christianity and the Environment.* Darton, Longman and Todd, London, 1990.

Buckley, Michael J. 'Religion and Science: Paul Davies and John Paul II'. *Theological Studies* 51/2, June 1990, pp. 310–24.

Byrne, Brendan. *Inheriting the Earth: The Pauline Basis of a Spirituality for Our Time.* St Paul Publications, Homebush, NSW, 1990.

Capra, Fritjof. *The Tao of Physics.* Tenth Edition, Flamingo, London, 1985.

The Turning Point: Science, Society and the Rising Culture. Flamingo, London, 1984.

Capra, Fritjof and David Steindl-Rast with Thomas Matus. *Belonging to the Universe. Explorations on the Frontiers of Science and Spirituality.* Harper San Francisco, New York, 1991.

Carmody, John. *Ecology and Religion: Toward New Christian Theology of Nature.* Paulist Press, New Jersey, 1983.

Carroll, Denis, 'Creation'. *The New Dictionary of Theology.* (eds.) J. Komonchak *et al*, Michael Glazier, Wilmington, Delaware, 1987, pp. 246–59.

Christiansen, Drew. 'Ecology, Justice and Development'. *Theological Studies*, March 1989, pp. 64–81.

Daly, Herman E. and Cobb, John B. *For the Common Good: Redirecting the Economy toward Community, Environment and a Sustainable Future.* Beacon Press, Boston, 1989.

Daly, Gabriel. *Creation and Redemption*. Gill and Macmillan, Dublin, 1988.

Davies, Brian. *The Thought of Thomas Aquinas*. Clarendon Press, Oxford, 1992.

Davies, Paul. *The Mind of God: Science and the Search for Ultimate Meaning*. Simon & Schuster, Sydney, 1992.

Doran, Robert. *Theology and the Dialectics of History*. University of Toronto Press, Toronto, 1990.

Edwards, Denis. *Jesus and the Cosmos*. St Paul Publications, Sydney, 1991.

'The Integrity of Creation: Catholic Social Teaching for an Ecological Age'. *Pacifica* 5, 1992, pp. 182–203.

Made from Stardust. Collins Dove, Melbourne, 1992.

Faricy, Robert. *Wind And Sea Obey Him: Approaches to a Theology of Nature*. SCM, London, 1982.

Fox, Matthew. *The Coming of the Cosmic Christ*. Collins Dove, Melbourne, 1988.

Sheer Joy: Conversations with Thomas Aquinas on Creation Spirituality. Harper, San Francisco, 1992.

Fox, Warwick. *Toward Transpersonal Psychology: Developing Foundations for Environmentalism*. Shambhala, Boston and London, 1990.

Franklin, Ursula. *The Real World of Technology*. CBC Enterprises, Toronto, 1990.

Gelwick, Richard. *The Way of Discovery: An Introduction to the Thought of Michael Polanyi*. Oxford University Press, New York, 1977.

Gilkey, Langdon. *Maker of Heaven and Earth: The Christian Doctrine of Creation in the Light of Modern Knowledge*. Doubleday, New York, 1959.

Gomes, Gabriel. *Song of the Skylark I-II*. University of America Press, Lanham, Maryland, 1991.

Griffiths, Bede. *A New Vision of Reality: Western Science, Eastern Mysticism and Christian Faith*. Fount, London, 1992.

Honner, John. 'Not Meddling with Divinity: Theological Worldviews and Contemporary Physics'. *Pacifica* 1/3, 1988, pp. 251–72.

'The New Ontology: Incarnation, Eucharist, Resurrection and Physics'. *Pacifica* 4/1, Feb. 1991, pp. 15–50.

Jaki, Stanley L. *The Purpose of It All*. Scottish Academic Press, Edinburgh, 1990.

Jastrow, Robert. *God and the Astronomers*. W. W. Norton, New York, 1978.

Johnson, Elizabeth A. *She Who Is: The Mystery of God in Feminist Theological Discourse*. Crossroad, New York, 1992.

Keen, Sam. *The Passionate Life: Stages of Loving*. Harper and Row, San Francisco, 1983.

Kelly, Tony (Anthony J.). *Trinity of Love: A Theology of the Christian God*. Michael Glazier, Wilmington, Delaware, 1989.

Touching on the Infinite: Explorations in Christian Hope. Collins Dove, Melbourne, 1990.

'Wholeness: Catholic and Ecological'. *Pacifica* 3/2 June 1990, pp. 201–23.

King, Thomas, M. *Teilhard de Chardin: The Way of the Christian Mystics.* Volume 6, Michael Glazier, Wilmington, Delaware, 1988.

LaChance, Albert. *Greenspirit: Twelve Steps in Ecologial Spirituality.* Element, Rockport, Massachusetts, 1991.

Lonergan, Bernard. *Method in Theology.* Darton, Longman and Todd, London, 1971.

Lovett, Brendan. *Life before Death: Inculturating Hope.* Claretian Publications, Quezon City.

Lyons, J.A. *The Cosmic Christ in Origen and Teilhard de Chardin.* Oxford, OUP, 1982.

MacDonagh, Sean. *To Care for the Earth.* Chapman, London, 1986,
The Greening of the Church. Geoffrey Chapman, London, 1990.

Mackin, Theodore, S.J. *The Marital Sacrament. Marriage in the Catholic Church.* Paulist Press, Mahwah, New Jersey, 1989.

McFague, Sallie. *Models of God: Theology for an Ecological, Nuclear Age.* SCM, London, 1987.

Macquarrie, John. *Principles of Christian Theology.* SCM, London, 1972.
Jesus Christ in Modern Thought. SCM, London, 1990.

Martelet, Gustave. *The Risen Christ and the Eucharistic World.* trans. René Hague, Seabury, New York, 1976.

Meeks, M. Douglas. *God the Economist: The Doctrine of God and the Political Economy.* Fortress Press, Minneapolis, 1989.

Midgley, Mary. *Science as Salvation: A Modern Myth and its Meaning.* Routledge, London, 1992.

Moltmann, Jürgen. *God in Creation: An Ecological Doctrine of Creation.* trans. M. Kohl, SCM, London, 1985.
The Way of Jesus Christ: Christology in Messianic Dimensions. SCM, London, 1990.

Mooney, Christopher F., S.J. *Teilhard de Chardin and the Mystery of Christ.* Collins, London, 1966.
'Cybernation, Responsibility and Providential Design'. *Theological Studies* 51/2, June 1990, pp. 286–309.
'Theology and Science: A New Commitment to Dialogue'. *Theological Studies* 52/2, June 1992, pp. 289ff.

Murphy, Charles M. *At Home on Earth: Foundations for a Catholic Ethic of the Environment.* Crossroad, New York, 1989.

Myers, N. ed. *Gaia: An Atlas of Planet Management.* Anchor, New York. 1984.

Nash, James S. *Loving Nature: Ecological Integrity and Christian Responsibility.* Nashville, Abingdon Press, 1991.

O'Murchu, Diarmid, M.S.C. *Religious Life: A Prophetic Vision.* Ave Maria Press, Notre Dame, Indiana, 1991.

Ormerod, N. 'Christianity and the Environment: Pursuing Truth through Metaphor'. *National Outlook* 12/4 June 1990, pp. 10–12.
'Renewing the Earth—Renewing Theology'. *Pacifica* 4 1991, pp. 295–306.

Peacocke, Arthur. *God and the New Biology*. J. M. Dent and Sons, London, 1986.

Percy, Walker. *Lost in the Cosmos: The Last Self-Help Book*. Arena, London, 1983.

Peters, Ted (ed.). *Cosmos as Creation: Theology and Science in Consonance*. Abingdon Press, Nashville, 1989.

Polanyi, Michael. *Personal Knowledge*. University of Chicago Press, Chicago, 1958.

Polkinghorne, John. *One World: The Interaction of Science and Theology*. SPCK, London, 1986.
Science and Creation: The Search for Understanding. SPCK, London, 1988.
Science and Providence: God's Interaction with the World. SPCK, London, 1989.

Pope, Stephen J. *The Contributions of Contemporary Biological Anthropology to Recent Roman Catholic Interpretations of Love*. Ph. D., University of Chicago, 1988.
'The Order of Love and Recent Catholic Ethics: A Constructive Proposal'. *Theological Studies* 55/2, June 1991, pp. 255–88.

Primavesi, Anne. *From Apocalypse to Genesis: Ecology, Feminism and Christianity*. Burns & Oates, London, 1992.

Principe, Walter, H. ' "The Truth of Human Nature" according to Thomas Aquinas: Theology and Science in Interaction' in *Philosophy and the God of Abraham: Essays in Memory of James A. Weisheipl, OP*. R. James Long (ed.), Pontifical Institute of Medieval Studies, Toronto, pp. 161–77.

Russell, R., W. Stoeger, G. Coyne (eds.). *Physics, Philosophy and Theology: A Common Quest for Understanding*. Vatican Observatory, Vatican City, 1988.

Santmire, H. Paul. *The Travail of Nature: The Ambiguous Ecological Promise of Christian Theology*. Fortress Press, Philadelphia, 1985.

Schumacher, E.F. *A Guide for the Perplexed*. Harper & Row, New York, 1977.

Schilling, Harold K. *The New Consciousness in Science and Religion*. United Church Press, Philadelphia, 1973.

Segundo, Juan Luis. *An Evolutionary Approach to Jesus of Nazareth*. John Drury (ed.) and trans. Orbis, Markynoll, 1988.

Sheldrake, Rupert. *The Rebirth of Nature: The Greening of Science and God*. Bantam Books, New York, 1991.

Smart, Ninian and Stephen Constantine. *Christian Systematic Theology in a World Context*. Marshall Pickering, London, 1991.

Stefano, Frances. 'The Evolutionary Categories of Juan Luis Segundo's Theology of Grace'. *Horizons* 19/1, Spring 1992, pp. 7–30.

Swimme, B. *The Universe is a Green Dragon: A Cosmic Creation Story*. Bear and Co., Santa Fe, 1984.

Tanner, Kathryn. *God and Creation in Christian Theology*. Blackwell, London, 1988.

Teilhard de Chardin, Pierre. *The Divine Milieu*. Harper & Row, New York, 1972.

Thompson, William Irving (ed.). *Gaia. A Way of Knowing: Political Implications of the New Biology*. Lindesfarne Press, Hudson, 1987.

Thornton, Lionel. *The Incarnate Lord*. Longmans, Green, London, 1928.

Tilby, Angela. *Science and the Soul: New Cosmology, The Self and God.* SPCK, London, 1992

Toolan, David S., S.J. '"Nature is a Heraclitean Fire". Reflections on Cosmology in an Ecological Age'. *Studies in the Spirituality of the Jesuits* 23/5 November 1991 [whole issue].

Toynbee, Arnold. *Mankind and Mother Earth. A Narrative History of the World*. Oxford University Press, New York, 1976.

Vale, Carol Jean. 'Teilhard de Chardin: Ontogenesis vs. Ontology'. *Theological Studies* 53/2, June 1992, pp. 313–37.

Voegelin, Eric. *Order and History IV: The Ecumenic Age*. Louisiana State University Press, Baton Rouge, 1974.

Westermann, Claus. *Creation*. trans. John Scullion, SPCK, London, 1974.

Wilber, Ken. *Up From Eden. A Transpersonal View of Human Evolution*. Shambhala, Boulder, 1983.

Wildiers, N.M. *The Theologian and his Universe: Theology and Cosmology from the Middle Ages to the Present*. Seabury, New York, 1982.

Wink, Walter. *Unmasking the Powers: The Invisible Forces That Determine Human Existence. The Powers*. Vol.2, Fortress, Philadelphia, 1987.

Winter, Gibson. *Liberating Creation*. Crossroad, New York, 1981.

Yates, John C. *The Timelessness of God*. University Press of America, Lanham, Maryland, 1990.

Zohar, Danah. *The Quantum Self: A Revolutionary View of Human Nature and Consciousness Rooted in the New Physics*. Bloomsbury, London, 1990.

Index of Names and Topics